The Genera of Myxomycetes

The Genera of Myxomycetes

G. W. MARTIN

C. J. ALEXOPOULOS

M. L. FARR

Illustrations by Ruth McVaugh Allen

UNIVERSITY OF IOWA PRESS

Iowa City

Library of Congress Cataloging in Publication Data

Martin, G. W. (George Willard), 1886–1971.
The genera of Myxomycetes.

Bibliography: p.
Includes index.
1. Myxomycetes. 2. Myxomycetes—Classification.
3. Fungi—Classification. I. Alexopoulos, Constantine
John, 1907– . II. Farr, Marie Leonore,
1927– . III. Title.
QK635.A1M37 1983 589.2′9 83–5092
ISBN 0–87745–124–9

University of Iowa Press, Iowa City 52242

Printed in the United States of America

To the memory of
THOMAS HUSTON MACBRIDE
1848–1934
and of
GEORGE WILLARD MARTIN
1886–1971

Contents

Note: Page numbers for the *Plates* follow the original pagination of the 1969 edition, for purposes of convenient reference.

Preface to the 1969 Monograph

In recent years there has been a marked revival of interest in the Myxomycetes. Improved methods of culture, in the form of 2-member cultures of a pure strain of a myxomycete of known origin combined with a known strain of bacteria and, as yet to a lesser extent, of the plasmodium alone on a nutrient medium, have resulted in a number of papers discussing the physiology, cytology, and variation in ploidy of both amoebae and plasmodia, and incidentally stressing the modifications which may appear under cultural conditions. These and other studies, added to the long-known variations which a single species may assume within the wide range of environmental influences occurring in natural habitats, make it clear that many of our bases for classification must be reconsidered. It is too early as yet to evaluate the full impact of this new information on the taxonomy of the group, but in the meantime a survey of the known species, even though they are arranged in a conservative framework, may serve a useful purpose as a summary of current taxonomic approach.

The authors are aware of many of the deficiencies of the present treatment. It seems quite certain that some forms here recognized as species will, when better known, be reduced to synonymy with other species, and that some common and widely distributed species which show wide variation will perhaps be split. At a higher level, some genera may come to be regarded as superfluous, others will be broken up, and the families, orders and higher categories will be modified and rearranged. This will always be true of a work of this kind. An attempt has been made to cite and comment upon some of the suggested changes which have already been proposed and, when it seems appropriate, to mention others which may be worthy of consideration.

The illustrations are based on sketches and specimens from the Iowa collection sent to Mrs. Allen, who supplemented such material by the study of specimens from her own collection and from other sources, notably the material in the collection at the Philadelphia Academy of Natural Sciences, including the Bilgram Collection, belonging to the Leidy Microscopical Society, also housed at the Academy.

We are grateful to the numerous correspondents who have sent us material for study. It is impossible to mention them all by name, but among those who have made noteworthy contributions are Dr. Wm. Bridge Cooke, Mme. N. E. Nannenga-Bremekamp, Dr. Marie L. Farr, Dr. Travis E. Brooks, Dr. Donald T. Kowalski, and Dr. K. S. Thind. To these, and to many others we extend our hearty thanks.

This work has been made possible by grants of the National Science Foundation: G-13426 and GB-3855 to the senior author, and in part by G-6382, G-19343, GB-248, GB-2738, GB-4657 and GB-6812X, to the junior author as included in the long-term support of his research program on the Experimental Approach to the Taxonomy of the Myxomycetes. For this assistance we are deeply appreciative.

G. W. Martin

January, 1969 C. J. Alexopoulos

Preface to the Present Edition

Since the publication of the 1969 monograph, *The Myxomycetes*, myxomycete research in all phases has continued to move forward at a rapid pace. Several books and review articles (listed in the text and bibliography) are now available covering cytology, genetics, physiology, and related topics concerning these organisms. The application of refined culture techniques and highly sophisticated instrumentation has vastly increased our knowledge and understanding of the structure and function of the various stages of the myxomycete life cycle. At the same time, these techniques made possible complete life cycle studies, in the laboratory, of a growing number of species in various orders. The resulting developmental studies have produced new concepts concerning the relationships among the higher taxa of the Myxomycetes, and led to some major changes and rearrangements in the classification system. The present taxonomic work, representing an abridged version of the Martin and Alexopoulos monograph, incorporates these changes, bringing the classification up to date to the genus level.

In the updated portions of the text, I endeavored to maintain the same academic level, a comparable proportion of nontaxonomic introductory material, and an adherence, as much as possible, to the style and spirit of the 1969 opus.

I am greatly indebted to C. J. Alexopoulos for encouragement and help in many ways; to M. Blackwell for critically reading the manuscript and supplying many references and suggestions that considerably improved the final product; and to H. C. Aldrich for reviewing the latter. In addition, I thank H. C. Aldrich and E. Haskins for supplying literature citations, and D. T. Kowalski, H. W. Keller, and the curators of the herbaria FH, K, NY, and TRTC for lending me needed specimens for study.

M. L. Farr

April, 1983

I. Introductory Remarks*

The Myxomycetes are fungus-like organisms characterized by an assimilative phase in the form of a free-living, multinucleate, acellular, mobile mass of protoplasm, the *plasmodium*, and a sporulating phase consisting of a mass of spores typically borne in a simple or complex, membranous or tough, noncellular spore case. Within the latter, in addition to the spores, there is often a system of free or netted threads forming a *capillitium*, or a *pseudocapillitium*. Certain groups also contain characteristic calcareous accretions within or without the spore case, or both.

The members of the group are widely distributed, occurring wherever conditions on the earth's surface permit growth of vegetation, but are particularly abundant in forested areas, where they appear in great profusion on dead and decaying wood, or woody litter, and on dead leaves. Not rarely the plasmodium creeps up the stems and over the leaves of living plants, and

*Pages *1–11* are based largely on a rough draft prepared by C. J. Alexopoulos, who kindly placed it at my disposal. M.L.F.

there sporulates. If such plants happen to be cultivated, some damage may be done, but such damage is incidental and there is no suggestion of true parasitism. On the other hand, the Myxomycetes cannot be described as saprobes. Characteristically they feed by phagocytosis on living bacteria and on fungal sporophores, spores, and mycelium as well as on bits of nonliving organic matter. Their nutrition is typically holozoic, but their myxamoebae and plasmodia are also capable of absorbing food in solution osmotically and by pinocytosis (Guttes and Guttes, **1960**). Certain species have been reported to cause damage in commercial mushroom beds (Harado, **1977**, and others). Koevenig and Liu (**1981**) have suggested that myxomycete plasmodia may be capable of breaking down cellulose.

As discussed later, there has been and still is much disagreement concerning the classification and relationships of the group. It was quite natural that early collectors, finding the true slime molds in the same situations and occurring with the fungi in their ordinary habitats, should have classified them with the fungi. Whether this was justified or not, it remains true that their collection and study has been mainly by mycologists, which fact has doubtless had some influence on many of the discussions as to their nature.

Classification is almost entirely based on the characters of the sporophores and their contents (spores, capillitium, and other structures). The characters of the plasmodia, however, are increasingly being employed to distinguish among subclasses and orders of the Myxomycetes, as indicated on pages *9–11*, *37*.

II.
General Life Cycle

The general life cycle of the Myxomycetes usually conforms to the following pattern:

1. The spores germinate into one or more (up to four) myxamoebae or flagellated cells (swarm cells or swarmers).

2. Myxamoebae multiply and give rise to a large population of cells. Mitotic spindles are open.

3. Compatible myxamoebae (or swarmers) fuse in pairs (syngamy) and give rise to zygotes.

4. Zygotes grow into multinucleate, acellular plasmodia by repeated synchronous mitotic divisions featuring spindles that are closed (i.e., within the nuclear envelope), without cytokinesis.

5. Plasmodia, under suitable environmental conditions, give rise to sporophores characteristic of the species, in which spores are formed.

6. Meiosis (in species studied with electron microscopy) occurs in the young (15–30 hr) spores and three of the four meiotic nuclei appear to disintegrate, resulting in uninucleate, haploid mature spores.

7. Under adverse conditions, the motile stages may change into resistant structures (microcysts, sclerotia).

Although this is believed to be the typical life cycle of the Myxomycetes, a number of variations have been discovered in certain species. For example, it is known that, at least in certain extensively analyzed strains, plasmodium formation is not necessarily dependent on cell or nuclear fusion, or on ploidy levels, but is apparently a differentiation process regulated by gene expression and control (see p. *14*). The details of life history and nuclear cycle are discussed in section IV beginning on page *13*.

III. General Morphology

The Sporophore — Mature fructifications (sporophores) occur in three distinguishable forms, as *sporangia*, as *plasmodiocarps*, and as *aethalia*. In the sporangial forms, one to many sporangia — in some cases thousands — develop nearly simultaneously from a single plasmodium. Generally the individual sporangium is of characteristic shape, size range, and color and in most species is rather consistently stalked or sessile. The sporangia of a cluster are seated on a typically horny, membranous sheath, the *hypothallus*, in some species very conspicuous and in others difficult to see or not visible. It is secreted by the plasmodium at the time of sporulation, or simply represents the plasmodial sheath. Lime, when present, may be in the form of amorphous granules or of highly characteristic crystals. It may be restricted to the stalk and its continuation within the sporangial cavity, the *columella*, which may arise from the base in sessile forms; it may be deposited in the hypothallus or the capillitium; it may form an outer broken or solid layer on the peridium, or it may occur in various combinations of these possibilities. The capillitium may be entirely limy, as in typical species of *Badhamia* or the lime may be aggregated in nodes connected by limeless tubules, as in *Physarum*. The nodes may be evenly distributed throughout the sporangium or they may be more concentrated in the center, forming a *pseudocolumella*. A plasmodiocarp is like a sporangium but is more or less elongated and often forms a network on the substratum (Pl. XII, Fig. 119a). It follows, in general, the lines of the major veins of the plasmodium from which it arose, often with breaks, and sometimes the breaks are so many that some of the segments are sporangium-like and are, indeed, sessile sporangia. Consequently the line between sporangiate and plasmodiocarpous sporophores is not always clear (Pl. XXVII, Fig. 250b), but in species which are characteristically plasmodiocarpous there is nearly always some indication of that fact even in fruitings that include a large proportion of sporangia. Plasmodiocarps are nearly always sessile; very rarely, when formed on the lower side of a branch or log they may be pendent by delicate threads (Pl. XXIV, Fig. 224a). In forming an aethalium, the strands of the plasmodium hold together in a tangled, pulvinate mass and while the inner portion proceeds to form spores exactly as in a sporangium, the outer tubes collapse and adhere together to form a crust, the cortex. Although the cortex is analogous with the peridium, there are differences in the development of the two structures, as shown in ultrastructural studies on the aethaliate genera *Lycogala* and *Enteridium* (Eliasson, **1981**) and a plasmodiocarpous species of *Perichaena* (Charvat et al, **1973**). In all Myxomycetes studied, however, formation of peridium as well as of cortex involves autolysis of peripheral protoplasm during fruiting body development (above-cited authors; Blackwell, **1974**). Under some circumstances aethalial development is checked and the final product resembles a densely packed cluster of sporangia, but this is rarely the case. A pseudoaethalium is a mass of closely compacted sporangia which simulate an aethalium, but in which the individual sporangia are clearly distinguishable at maturity (Pl. II, Figs. 17a, 18a, b, 19a) or, in the special case of *Dictydiaethalium* (Baker, **1933**), almost to maturity. Here too, intermediate forms can present problems of identification.

Cohen (**1942**), studying the variability of the fructifications of *Didymium squamulosum, Physarum polycephalum,* and *Badhamia foliicola,* reached the conclusion that the type of fructification formed is probably dependent on the surface tension of the plasmodium. A low surface tension would tend to produce plasmodiocarps, whereas a higher surface tension would favor the formation of sessile, spherical sporangia. A stalked sporangium would result when material is secreted preferentially at the substrate-plasmodium interface. However, it is undeniable that some forms, such as *Hemitrichia serpula,* seem always to produce plasmodiocarps, whereas others, such as species of *Stemonitis,* never do. Nor is Cohen's explanation entirely in accord with laboratory and field observations mentioned later.

In the endosporous Myxomycetes, a peridium typically surrounds the spore mass. The peridium may be delicate or tough, covered or impregnated with lime or other substances, and consist of from one to three remote or appressed layers. At maturity it may open by a predetermined lid, split irregularly or along definite lines of dehiscence, or slough off in various ways, some of which are sufficiently characteristic to be useful in species determination. In some genera the basal portion persists as a cup- or disk-shaped *calyculus*. In a number of species the peridium evanesces very early in the formation of the sporophore (Alexopoulos, **1960**a; Goodwin, **1961**) while the spores are being developed, or possibly is not formed at all, at least in the upper part of the sporangium, as reported for *Arcyria cinerea* (Mims and Rogers, **1975**), so that the mature spores are in no way covered. In the so-called exosporous types, the spores appear to be produced on the outside of a columnar, poroid, or morchelloid fructification (Pl. I, Figs. 1a, 2a, 3a).

One fact should be stressed. The plasmodium may develop its characteristic sporulating stage in less than 24 hours. If this occurs under conditions which cause unduly rapid drying or if repeated rains check the process, great variation may be induced. Under such influences, species which ordinarily have stalks may be sessile or nearly so, or the stalks may be inordinately long; sporangiate species may form plasmodiocarps; aethaliate forms may approach the sporangiate type; the characteristics and disposition of limy secretions may be altered; capillitial formation may be suppressed to varying degrees; spore maturation may be checked, resulting in sporelike bodies which are much larger than fully matured spores. Cold weather, and particularly frosts, may induce similar alterations. Experimental half-hour exposures to 37°C in the laboratory (heat shock) produced "abnormal" sporangia when applied during early stages of sporangium formation (Sauer et al, **1969**). Such variations are in large part responsible for the extensive synonymy found in the group. Great caution is indicated in describing as new, specimens which are the result of such environmental influences. They are not "abnormal"; they are the natural response of the organism involved to particular stimuli and must be so regarded. Giving them taxonomic status as named varieties serves only to complicate the nomenclature and to extend the meaning of the category "variety" beyond its legitimate significance. The same caution should be exercised in naming varieties based on sporophore color, which is highly variable in some species.

Capillitium and pseudocapillitium. A true capillitium, when present, is formed just before spore delimitation begins. The capillitium in mature sporophores may range from a few threads to a dense network; the threads may be solid, hollow, smooth or sculptured, and partly to entirely lime-filled. The capillitium may arise from a system of tubules formed by anastomosing vacuoles within the protoplasm of the ripening sporophore, as, for example, in *Trichia fallax* [=*T. decipiens*] (Strasburger, **1884**), *Hemiarcyria* [=*Hemitrichia*] *clavata* (Harper and Dodge, **1914**), *Physarum polycephalum* (Howard, **1931**), and *Arcyria cinerea* (Mims, **1969**). In some other noncolumellate species such as *Badhamia gracilis* (Welden, **1955**) and *Perichaena vermicularis* (Charvat et al, **1974**), the capillitium develops partly in this manner and partly from invaginations of the plasma membrane inside the peridium. In columellate species such as *Didymium iridis* (Welden, **1955**) and species of *Stemonitis* (Bisby, **1914**; Ross, **1958**b, **1961**a; Indira, **1971**; Mims, **1973**), a portion of the capillitium originates at the columella, while the remainder is formed from tubular vacuoles or by invaginations of the peridial membrane (depending on the species), the two systems anastomosing and eventually joining. The capillitial strands in *Comatricha* (Goodwin, **1961**) and *Lamproderma* (Ross, **1958**b) are extensions of the strands which form the stalk and columella and which bend outward, grow, and branch toward the peridium. In the Physaraceae, as previously stated, the capillitium may be entirely limy or may consist of a system of hyaline tubules supporting calcareous nodes. In the Didymiaceae lime is rarely present in the capillitium even though it is characteristically deposited in other parts of the sporophore. Even in *Physarum*, the amount of lime present in the capillitium and the peridium seems to be dependent at least to some extent on the environment, limeless or nearly limeless fruitings occurring occasionally in many species which typically secrete lime. It must be emphasized here that "lime" refers to calcium carbonate which is visible with the unaided eye or through a light microscope. Recent studies utilizing energy dispersion X-ray spectroscopy have revealed the presence of the element calcium in considerable quantities in some so-called nonlimy genera (Schoknecht, **1975**; Nelson et al, **1977**). Furthermore, the presence of calcium oxalate in the form of deposits on the peridium of some species of *Perichaena* has been known for a long time (Lister, **1925**), and was more recently confirmed by Schoknecht and Keller (**1977**).

The chemical composition of the capillitium has not been determined with certainty. Chitin has been reported to be present in the capillitium of *Stemonitis fusca* (Cihlar, **1916**), and of *Hemitrichia serpula* (Locquin, **1947**). But in view of the inability of many workers (Ulrich, **1943**; Goodwin, **1961**) to demonstrate its presence in the majority of species which have been investigated, a reexamination of the whole problem using modern techniques is indicated.

In a number of species the capillitium is so constructed as to be a definite aid in the dissemination of the spores. In such species as *Arcyria nutans*, *A. oerstedtii*, *Hemitrichia clavata*, etc., it is elastic, expanding greatly when the peridium ruptures and shedding large quantities of spores. In

Trichia favoginea the elaters which form the capillitium are hygroscopic, twisting and untwisting according to the humidity of the surrounding atmosphere (Ingold, **1939**). The spores are thus stirred and some may be forcibly expelled by such capillitial action.

A pseudocapillitium occurs only in aethalia or pseudoaethalia, and usually represents empty strands of the plasmodium from which the protoplasm has been evacuated before spore formation. In *Dictydiaethalium* it is formed from the thickened parts of the sporangial walls, where the individual sporangia have been pressed together, and which remain after the thinner parts of the wall have disappeared. In other genera it may occur in the form of threads, bristles, membranes, or perforated plates. If threadlike, its threads vary in width at different points and are distinctly irregular (Pl. II, Figs. 16b, 23c; Pl. III, Figs. 27b, 30b). Olive (**1975**) suggested that a different term should designate the structures in *Dictydiaethalium*, since they are most likely not homologous to the pseudocapillitium of other genera. The bristles in *Tubifera bombarda* may belong in the same category with the threads of *Dictydiaethalium*. In some genera, such as *Fuligo*, both a true capillitium and a pseudocapillitium are present in the mature aethalium.

The Spores. The spores are nearly always globose although in a few species they are predominantly oval in shape (Pl. XXIII, Fig. 212). In color they may vary by transmitted light from hyaline to almost black, and ranging through various shades of violet, pinkish, yellow, and brown. In the darker spores there is often a strong purplish tinge to the color. In mass the color is more obvious, particularly in the paler forms. Many are brilliant yellow, pink, ferruginous, or purple. Even in the very dark-spored species, where the spore mass is black, there is often a distinct purplish cast. In a few species, particularly in *Badhamia* but also in other genera, the spores are aggregated into characteristic clusters (Pl. XIII, Fig. 214c, d). The spore wall may appear perfectly smooth or it may be echinulate, verrucose, or elaborately reticulate. The spore wall of some species of *Echinostelium* exhibits characteristic thickenings at points of contact with other spores (Pl. VI, Figs. 57b, 58b, 60c). Within certain limits these characters, as well as size, are remarkably constant and serve as useful means of identification.

Two wall layers may be easily distinguished in all myxomycete spore walls examined by transmission electron microscopy. The inner layer, next to the plasmalemma, is electron transparent and usually very much thicker than the electron dense outer layer. The latter bears the ornamentations (spines, warts, or ridges) if any. Whether these constitute a third layer or are an integral part of the outer layer is a matter of interpretation of the electron micrographs. In the species studied, at least some of the ornamentations are laid down before the other wall layers (Aldrich and Blackwell, **1976**).

Noteworthy ultrastructural components of young spores are dense bands of microtubules of as yet undetermined function; a large central digestive vacuole (before meiosis), smooth ER, Golgi vesicles, and food reserve in the form of glycogen particles. Centrioles are formed in the ripe spores (Aldrich and Blackwell, **1976**).

Little is known about the chemical composition of the spore wall. Schuster's (**1964**) electron micrographs of *Didymium nigripes* spores show two wall layers, of which the inner coat gave a cellulose reaction and the outer may consist of chitin, according to that author. In accordance with the findings of McCormick et al (**1970**), the spore walls of *Physarum polycepha-*

lum resemble spherule walls, consisting mostly of galactosamine polymer (81%), glycoprotein, and about 15% melanin, with small amounts of amino acids and phosphate. Henney and Chu's (1977) experiments showed galactosamine likewise to be the major constituent in microcyst walls of *Physarum flavicomum*. The slime sheath was found to consist of a galactose-containing polysaccharide and a protein similar in amino acid composition to those in the microcyst and microsclerotial walls. The predominance of galactosamine represents a striking difference in wall structure from that of the fungi, cellular slime molds, and Protozoa so far studied in this respect (McCormick et al, **1970**).

Swarm Cells and Myxamoebae. When a spore germinates, usually one, but occasionally as many as four, swarm cells or myxamoebae emerge. The immediate environment of the spore at the time of germination determines to some extent whether the emerging protoplast will be flagellated or not. In water or a weak nutrient solution the spores often release flagellated protoplasts, which may remain in this condition from a few hours to many days, apparently depending on the species (Ross, **1958**a). On a moist surface devoid of free water they release amoebae. In some species the entire life cycle may be completed without any flagellate stage if water is withheld throughout (Alexopoulos, **1960**a). Similarly, certain variant strains of *Physarum flavicomum* lack both the swarm cell and myxamoebal stages from their life cycles and proceed to develop plasmodia directly from the protoplasts liberated from the germinating spores without the formation of either swarmer or myxamoebal populations (M. R. Henney, **1967**). What happens in nature is of course entirely unknown.

For a long time the swarm cells of the Myxomycetes were believed to be uniflagellate with a single anterior whiplash flagellum. Elliott (**1949**) in a study of 59 species showed that the swarm cells typically possess two anterior whiplash flagella one of which, in many species, is much shorter than the other, is recurved, and because it may be appressed against the protoplast of the swarmer, is difficult to demonstrate. Locquin (**1949**) using cinephotomicrography and phase-contrast optics independently found a short, immobile flagellum at the base of the long active one in all 58 species he studied. Eight years later Ross (**1958**a) stated that the apparently uniflagellate cells of 19 species he studied possess a second, extremely short flagellum which is usually appressed to the cell membrane of the swarm cell. Cohen (**1959**) reopened the subject of flagellation at the Ninth International Botanical Congress in 1959. His electron micrographs of whole swarm cells showed but a single flagellum in many of the swarm cells examined. Cohen demonstrated the presence of flagellumlike pseudopodia originating at the base of the true flagellum and expressed the belief that these structures may have been mistaken for flagella by other workers. He did not deny that some species or strains typically may possess biflagellate swarm cells, but insisted that flagellation is not the same for all species.

Thin sections of swarm cells of *Stemonitis nigrescens*, *Comatricha laxa*, *Physarum flavicomum*, and *Didymium iridis* examined by Aldrich (**1968**) under the electron microscope unmistakably revealed a second short, recurved flagellum. Thus, in spite of conflicting reports there is very convincing evidence, both from the optical and the electron microscope, indicating that myxomycete swarm cells are typically biflagellate. N. S. Kerr (**1960**), in a study of flagellation of *Didymium nigripes*, made the very interesting observation that "Newly flagellated cells at first possess a single flagellum, 10 μ long. Cells which have been flagellated for several hours often have 2 flagella." This may be a partial explanation for the apparently discordant reports. Nearly all species studied with the electron microscope have a pair of centrioles (Schuster, **1965**; Aldrich, **1968**; Mims, **1971**; Nelson and Scheetz, **1975**). Haskins (**1978**a) has shown that, in *Echinostelium minutum*, swarm cells may be equipped with from one to four (rarely up to eight) flagella, but possess at least two pairs of centrioles each. Perhaps the most accurate statement of the situation as we know it at present (Alexopoulos and Mims, **1979**) is that myxomycete swarm cells are potentially anteriorly biflagellate, but that uniflagellate and multiflagellate cells are sometimes produced.

It has also been demonstrated (Schuster, **1965**; Aldrich, **1968**) that myxomycete flagella, in common with those of most other organisms, have the 9 + 2 microtubule pattern.

Indira (**1964**) reported that the plasmodium of *Arcyria cinerea* when placed in water may cut off swarm cells either from the advancing margin or from the veins, and submitted evidence that this may also be done by the plasmodia of *Physarum compressum* and *Stemonitis herbatica*. It was not determined whether such swarm cells are haploid or diploid. Ross and Cummings (**1967**), who observed a similar phenomenon in two cultures of *Physarum ?pusillum*, believe it to represent abnormal behavior.

The employment of modern techniques such as phase-contrast optics, cinephotomicrography, and electron microscopy has made it possible to observe the structure and behavior of the swarm cells better than ever before. Flagellumlike pseudopodia have been seen to originate at the base of the flagella and migrate toward the posterior end (Cohen, **1959**; Koevenig, **1961**); streaming cytoplasm in the swarm cell is now easily observed; and the stickiness and pseudopodial activity of the posterior end have been adequately demonstrated. The swarm cells are uninucleate, with the nucleus in the anterior end just below the basal bodies. In addition to the nucleus and centrioles, Golgi complexes, mitochondria, food and contractile vacuoles and rough ER occur in the cytoplasm of swarmers and myxamoebae (Schuster, **1965**; Aldrich, **1968**). The nuclear envelope breaks down during late prophase and polar centrioles appear (Olive, **1975**); mitosis is followed by cytokinesis. This situation is in contrast to plasmodial mitosis (p. *14*).

A noteworthy recent development is the *in vitro* synthesis of myxomycete microtubules by heat-induced polymerization of microtubule protein from myxamoebae of *Physarum polycephalum* (Quinlan et al, **1981**). These researchers obtained similar results from *in vitro* and *in vivo* tests of susceptibility to various microtubule inhibitors, and confirmed that the pattern of drug sensitivity in *Physarum polycephalum* differed from that of higher organisms.

Under adverse conditions, myxamoebae become transformed into dormant microcysts, a state suggested to be highly significant in the survival and dissemination of the species (Aldrich and Blackwell, **1976**; Collins, **1979**). Ultrastructural studies show the microcyst walls to be single-layered in most of the few species examined, and composed of randomly arranged microfibrils (Collins, **1979**). H. R. Henney and Chu (**1977**) found the microcyst walls of *Physarum flavicomum* to consist mostly of polysaccharide (galactosamine and other sugars), lipids, and protein. In *Didymium nigripes* (Schuster,

1965) and *D. iridis* (Aldrich and Blackwell, **1976**) the microcyst walls are spinulose. According to the latter authors, microcysts can remain viable for long periods of time on agar slants and in lyophilized state.

The plasmodium. Plasmodia are commonly seen in the field although, by reason of their way of life, less frequently than the fructifications. Those readily observed usually belong to the Physarales. They are in the form of a network of gelatinous veins, fanning out toward a continuous advancing margin. Plasmodia may be hyaline, white, yellow, violet, red, or nearly black with various intermediate tints and shades. The color appears to be reasonably constant for a given species but is affected by acidity or alkalinity, to some extent by light and temperature, and very commonly by material, including food, taken from the substratum. The plasmodium of *Physarum polycephalum* is reported to have a pleasant smell (Dee, **1975**). In most species, the plasmodium is covered by a slime layer containing microfibrils (Alexopoulos and Mims, **1979**). This layer is shed behind the creeping plasmodium in the form of readily visible "slime tracks" on the substrate, and is said to consist of galactose polymers and a glycoprotein (Simon and H. R. Henney, **1970**).

In a few species, of which *Physarum polycephalum* and *Physarella oblonga* are conspicuous examples, the plasmodium may spend practically all of its life at the surface, feeding upon the hymenial elements of fleshy fungi and associated bacteria. It is also capable of absorbing liquid food, as well as ingesting swarmers and myxamoebae of its own species (cannibalism). Further growth may occur by fusion with other genetically compatible plasmodia. In a large number of species, represented by *Ceratiomyxa fruticulosa, Physarum cinereum*, various species of *Stemonitis*, the Diacheas, and many plasmodiocarpous species, the plasmodium spends most of its time within the substratum, especially soil and decayed wood, apparently feeding on bacteria and other microorganisms and quite probably on fungus hyphae, and decayed wood, coming to the surface only when ready to sporulate. In many cases, however, it is found on the moist, dark underside of its substrate (Alexopoulos, **1973**). Little is known about the plasmodial habits of a very large group of Myxomycetes which includes most of the Liceas and most of the Cribrarias. It may be that in this group the plasmodium as a whole never appears on the surface, but that the primordia of what are to be individual sporangia emerge as droplets and proceed to maturity singly, but often in large numbers. On the other hand, it is possible that in many of these species, especially those with minute sporophores, the plasmodium may be formed on the surface, remain microscopic throughout its existence, and then form but a single sporangium at the time of sporulation. This is certainly the case in three species of *Echinostelium* (Alexopoulos, **1960**a, **1961**), in *Clastoderma debaryanum* (McManus, **1961**), and in *Protophysarum* (Blackwell, **1974**).

Plasmodia are admirably adapted to penetrating extremely minute pores. Moore (**1933**) has shown that the plasmodium of *Physarum polycephalum* may be induced to pass through the pores of a Berkfeldt filter, although it cannot be forced through. More recently, Clark and Hakim (**1980**) found that diploid plasmodia of *Didymium iridis* passed through filters with 3 μm pore diameter, but tetraploid plasmodia did not.

Structurally there are at least three types of plasmodia (Alexopoulos, **1960**b). The most common, best known, and most generally encountered type is the *phaneroplasmodium*, characteristic of the order Physarales. This takes the form of a robust, easily observed network of veins terminating in a fanshaped sheet of protoplasm with a definite margin. Examination of the veins under a low power of the microscope will show that each vein consists of an outer layer in which no streaming occurs and an inner core of streaming protoplasm. The protoplasm flows in one direction for about sixty to ninety seconds, the flow stops, and then the protoplasm flows in the reverse direction. In the phaneroplasmodium, the protoplasm is very granular, the inner and outer layers of the veins distinct, the advancing, fleshy fans conspicuous, and the margin definite.

A second type of plasmodium, the *aphanoplasmodium*, is known only from a few species of *Stemonitis, Comatricha, Lamproderma*, and one species of *Amaurochaete* (Farr, **1982**), but most of the Stemonitaceae may be expected to conform to this type. In its actively growing stage it is hyaline and inconspicuous. It consists of a very fine network of hyphalike strands in which the protoplasm is much less granular than that of the phaneroplasmodium. In artificial culture a definite margin and advancing fan are absent in the early stages of development, the hyphalike strands often continuing unbranched for a considerable distance and terminating either abruptly, or with the formation of many short branches or vesiclelike projections. Eventually, a definite margin is formed, but seldom the sheetlike fans with streaming channels so characteristic of the phaneroplasmodium. Only in the largest strands is the differentiation of inner and outer layer conspicuous, most of the veins exhibiting only a thin membrane which holds the reversibly streaming protoplasm within bounds. This type of plasmodium is never observed in nature until it masses just before differentiating. Whether it spends the major portion of its existence inside the substratum, as is generally thought, and comes to the surface at the time of sporulating, or whether it lives on the surface but because of its structure and hyaline character is invisible, is an interesting question.

The third plasmodial type, the *protoplasmodium*, is probably the most primitive. It is known from six species of *Echinostelium* (Whitney, **1980**), several species of *Licea*, and *Clastoderma debaryanum*, and probably will be found in many more minute species when they are studied in culture. Some minute myxomycetes, however, are reported to have true, if small, phaneroplasmodia; examples are *Protophysarum phloiogenum* (Blackwell and Alexopoulos, **1975**), *Didymium eremophilum* (Blackwell and Gilbertson, **1980**), and *Licea fimicola* (Keller and Anderson, **1978**). A protoplasmodium retains juvenile characteristics throughout its life. It never grows larger than a millimeter in diameter. It is highly granular, but forms no veinlike strands, reticulum, or advancing fans, although some differentiation may take place during its migration (Haskins, **1978**b). Its protoplasm streams slowly, inconspicuously, and irregularly. At the time of sporulation, the protoplasmodium typically produces but a single, minute sporangium.

A plasmodial type which combines some of the characters of the phanero- and the aphanoplasmodium has been described by Alexopoulos (**1960**b) and later by McManus (**1962**), Indira and Kalyanasundaram (**1963**), Ross (**1967**b), and Rammeloo (**1976**). This type seems to be characteristic of some of the Trichiales. In this connection Lakhanpal and Mukerji's (**1981**) report of a protoplasmodium for *Calomyxa metallica* certainly is of interest and would bear further investigation. As the plasmodial stages of other Myxomycetes are studied more intensively, other types will undoubtedly emerge, but probably the three types described above form centers of evolution from which other types may have arisen.

Regardless of type, the plasmodium of the Myxomycetes contains many nuclei, but no trace of cell walls. It has been interpreted as a single multinucleate organism on the one hand and, on the other, as a multicellular organism in which the cell walls have disappeared. Luyet (**1940**) pointed out that both of these explanations are attempts to fit the recognized facts into the cell theory, in the belief that the theory must encompass all living things. Martin (**1932, 1957**, and in various other publications) has suggested that Dobell's term "non-cellular" should probably apply to the myxomycete plasmodium. This is not to deny that the spores, the swarm cells or motile units which arise from them, and the zygotes conform to the usual definition of cells.

In addition to the uninucleolate nuclei, the plasmodial endoplasm contains mitochondria with tubular cristae, smooth and rough ER, ribosomes, vacuoles, and sometimes microfibrils, microtubules, pigment granules, as well as particles and inclusions of unknown function (Dee, **1975**; Olive, **1975**).

The sclerotium. Under the influence of desiccation, adversely high or low temperature, lack of food, low pH, high osmotic pressure, sublethal doses of heavy metals, and probably other unfavorable conditions, the phaneroplasmodium may become transformed into a hard, horny resting stage, the *sclerotium*. Within the sclerotium, the protoplasm becomes aggregated in small masses or *macrocysts* ("spherules" when formed in liquid culture), each surrounded by a membrane (Jump, **1954**). This state may be induced in the laboratory by transfer to a non-nutrient salt solution or by addition of mannitol to the growth medium (Gorman and Wilkins, **1980**). Macrocysts vary in size from 10 to 25 μm in diameter, and in number of nuclei from none to about 14. They appear to be formed by the fusion of vesicles already present in the plasmodial protoplast. Upon fusion, the vesicles contribute their membranes to the new units formed (Stewart and Stewart, **1960**). Spherule formation is accompanied by polysaccharide slime production and involves changes in carbohydrate and protein metabolism as well as enzyme activity (Hüttermann, **1973**b). On the ultrastructural level, Golgi complexes appear during the transitional state (Hüttermann, **1973**b), and the smooth ER changes to rough ER (Goodman and Rusch, **1970**).

Sclerotia, when properly stored, retain their viability for one to three years and give rise to typical plasmodia when revived under favorable conditions. Some species overwinter in the sclerotial state and sclerotia collected outdoors in the winter have been reported to revert to the plasmodial state in the laboratory (Alexopoulos and Mims, **1979**).

Aphanoplasmodia do not develop hard, horny sclerotia. Instead, when dry conditions set in, the plasmodial veins contract and separate into microscopic droplets which encyst and form a discontinuous pattern replicating that of the plasmodial strands. Such small cysts, the aphanosclerotia, are invisible to the unaided eye. Sporangia of certain species in the Stemonitaceae appear on bark placed in moist chambers within 24 hours after the bark is wet (Alexopoulos, **1964**a). The most logical explanation for such rapid development is the probable presence on the bark of aphanosclerotia which, upon hydration, become reconstituted into plasmodial droplets which proceed to sporulate almost immediately.

Protoplasmodia encyst in their entirety, each forming a cyst. Both aphanoplasmodia and protoplasmodia sometimes sclerotize under water for unknown reasons. In that event they remain encysted and cannot be induced by any known method to resume their activities.

Sporophore development. There appear to be two major types of sporophore development in the endosporous Myxomycetes. This was first noted by A. de Bary who described them in his classic textbook "Comparative Morphology and Biology of the Fungi, Mycetozoa, and Bacteria" (English translation, **1887**). Bary's description went largely unnoticed, however, until recent years when, one might say, these two types of development were rediscovered (Ross, **1958**b, **1961**a; Alexopoulos, **1969**; Mims, **1973**; Blackwell, **1974**) and made the basis for a modern classification of the Myxomycetes (Ross, **1973**; Alexopoulos, **1973**). Bary did not apply names to these two types of differentiation, but pointed out that one type is characteristic of *Stemonitis* and the other is common among the Liceales, Physarales (his Calcarinae) and Trichiales. Partially on the basis of type of sporophore development, Ross (**1973**) classified the endosporous Myxomycetes into two subclasses: the Myxogastromycetidae with a myxogastroid (subhypothallic) development and the Stemonitomycetidae with a stemonitoid (epihypothallic) development. At G. W. Martin's suggestion, Alexopoulos (**1969, 1973**) employed the terms subhypothallic and epihypothallic to these two types of development, respectively. Other authors, however, (Mims, **1973**; Blackwell, **1974**) rejected these terms as awkward and substituted "stemonitoid" for epihypothallic and "nonstemonitoid" for subhypothallic. Because "nonstemonitoid" is a negative term, Alexopoulos (in lit.) proposed the word "myxogastroid" for it. Thus the terms subhypothallic, nonstemonitoid, and myxogastroid are synonymous and all three may be found in the recent literature. Olive (**1975**) does not accept the two subclasses, one reason being the probable presence of a third, protostelid type of development in at least one species of *Echinostelium*.

In the myxogastroid type of sporophore differentiation, characterized by *Physarum polycephalum* (Guttes et al, **1961**), the plasmodium becomes concentrated in several places to form hemispherical mounds which become separated by the resorption of the connecting protoplasmic strands by the mounds. The slime layer of the plasmodium forms the hypothallus, if any. These mounds now elongate into tiny pillars (sometimes called papillae), each of which becomes transformed into a stalked, single or compound sporangium. In *Protophysarum phloiogenum*, and possibly other members of the Myxogastromycetidae, food vacuoles come to lie in the central portion of the plasmodium at the time it is concentrating to form mounds. As the papillae develop, the food vacuoles accumulate dense, fibrous material within them. Then they coalesce into a central vacuole and extrude their contents. Their remnants finally form the central core of the stalk which is surrounded by a stalk tube of similar fibrous material (Blackwell, **1974**). A fibrous sheath is also formed in the stalks of *Arcyria cinerea* (Mims and Rogers, **1975**).

In the stemonitoid type of sporophore development, characterized by the genera *Comatricha* (Jahn, **1899, 1931**; Goodwin, **1961**), *Lamproderma*

(Ross, **1961**a), and *Stemonitis* (Mims, **1973**), the plasmodium deposits a hypothallus on the substratum and then becomes concentrated into one or more, more or less spherical masses inside which it begins depositing a stalk on the hypothallus. As the stalk elongates, the protoplasm crawls upward and continues depositing material, always internally, to the tip of the stalk until the total height, predetermined by the inheritance of the slime mold and influenced by the environmental conditions prevailing at the time of sporulation, is reached. At this time, the protoplasmic sphere, now engulfing the tip of the stalk, may secrete a thin wall (the peridium) around it. Capillitial threads then begin to extend from the columella toward the surface of the developing sporangium. A portion of the capillitium also may arise from a system of tubular vacuoles in the protoplasm (Mims, **1973**).

IV. Life History and Nuclear Cycle

The endosporous species. The life history of all endosporous Myxomycetes which have been investigated follows in its gross aspects the same general pattern with only minor deviations at certain points by individual species. The spores germinate and release one to four uninucleate, haploid protoplasts which develop flagella or become amoeboid, according to moisture conditions, as explained on pp. *7* and *17*. In the amoeboid phase, in the presence of bacteria which may be used for food, the protoplasts may undergo successive divisions and build up large populations. In the absence of sufficient food, the myxamoebae round up and encyst. If encystment takes place in water, little can be done to cause excystment. If, however, encystment takes place on a relatively dry surface, the addition of an aqueous bacterial suspension often induces germination of the cysts and subsequent formation of flagellated cells which resume the life cycle. In the presence of free water, many myxamoebae change into flagellated cells which continue to feed.

When a certain critical concentration of cells has been reached, compatible amoeboflagellates behave as gametes and copulate in pairs, the swarm cells making contact at their sticky posterior ends. Whether copulation of amoebae or flagellated cells is characteristic of individual species is not certain. Ross (**1958**a) believed that it may be, but N. S. Kerr (**1961**) and Koevenig (**1961**) have shown for *Didymium nigripes and Physarum gyrosum*, respectively, that either pattern may prevail. It appears that contact of compatible cells for several hours is necessary before fusion actually takes place. To explain this delay, Ross and his coworkers (see Ross, **1979**) have postulated the requirement of an "induction period" which prepares compatible cells for recognition and fusion by mediating cell surface changes.

Heterothallism has now been adequately demonstrated in 14 of the 38 species which have been studied in this respect (Collins, **1979**). About half of the 14 species also encompass nonheterothallic forms. Heterothallism appears to be essentially of the bipolar type with multiple alleles at the mating type locus known in all the heterothallic species which have been investigated (Collins, **1963**; Dee, **1966**; Henney and Henney, **1968**). In *Physarum polycephalum*, however, two mating type loci have been found, one apparently controlling amoebal cell fusion and the other, plasmodium formation (Youngman et al, **1979**; Kirouac-Brunet et al, **1981**). It is of interest to note here that the gametes of the two mating types in *Didymium iridis* and *Physarum pusillum* were reported by Therrien (**1966**) to have different amounts of DNA in their nuclei. The DNA sum of the two mating types was almost exactly equal to the amount found in plasmodial nuclei. According to Collins (**1979**), 31 of the 38 species studied are nonheterothallic or include nonheterothallic isolates. Homothallism has generally been assumed if monosporous cultures yield plasmodia and sporulate, without any confirmation that nuclear fusions actually take place in such cultures. Collins (**1979**) correctly substituted the term "nonheterothallic" for forms evidently not possessing mating types. Plasmodia, however, can also arise apogamically (N. S. Kerr, **1967**; Dee, **1975**; Honey et al, **1981**) by mutations at the mating type locus. Nuclear fusions were not found to be prerequisite to plasmodium formation. It appears from the available data that at least some species, such as *Physarum polycephalum* (Dee, **1960**; Wheals, **1970**) and *Didymium iridis* (Collins, **1976**; Therrien and Yemma, **1974**; Therrien et al, **1977**; Gorman and Wilkins, **1980**) consist of homothallic, apogamic, and heterothallic strains. In other species which have been studied, such as *Fuligo cinerea* (Collins, **1961**) and *Physarum flavicomum* (M. R. Henney, **1967**), only homothallic (apogamic?) *or* heterothallic strains have been discovered so far. For a more detailed discussion on reproductive systems, consult Collins (**1979**) and Collins and Betterley (**1982**).

Once the zygote (or the incipient plasmodium in apogamic forms) is formed, it feeds and grows. Karyokinesis without cytokinesis takes place successively and transforms the cell into a plasmodium. Nuclear divisions in the zygote and plasmodium are noncentric (closed), exhibiting intranuclear spindles lacking polar centrioles, and a persistent nuclear membrane, as contrasted with the centric (open) nuclear divisions in the myxamoebae, in which the nuclear envelope breaks down during prophase and centrioles are present at both poles. In the plasmodium the divisions are synchronous and under rigidly controlled conditions the synchrony is absolute. DNA synthesis begins immediately after mitosis and lasts for one to two hours (Nygaard et al, **1960**). Olive (**1975**) reviews current information on DNA and RNA synthesis in considerable detail. It is of interest that the last protein essential to mitosis is transcribed only 15 min before metaphase. When microplasmodia (minute plasmodia, less than 1 mm in diam and containing only a few hundred nuclei) of *Physarum polycephalum*, each with a different mitotic cycle, are permitted to fuse, the mitosis is synchronized throughout the resulting plasmodium after an adjustment period of about 7½ hours. This represents about one-half of the interphase period. Minute plasmodia fuse one with another and with zygotes with which they come in contact, thus enlarging by accretion as well as growth.

The subject of heterokaryosis in the Myxomycetes is a very interesting one. The question is whether plasmodia with nuclei of different constitution are able to fuse. The few experiments which have been conducted to answer this question argue against the wide-spread occurrence of heterokaryosis. In those species examined, from two to thirteen loci were found to determine plasmodial compatibility (Carlile, **1973**). It has been demonstrated in the laboratory that certain genetically incompatible plasmodia are capable of fusing, but in such unions the heterokaryotic condition is unstable (Poulter and Dee, **1968**), or they produce a gene-controlled cytotoxic or lethal effect on one of the partners in *Physarum polycephalum* (Carlile and Dee, **1967**), *Badhamia utricularis* (Carlile, **1974**), *Physarum cinereum* (Clark, **1977**), and *Didymium iridis* (Clark and Collins, **1973**; Ling and Clark, **1981**). According to Ling and Clark (**1981**), somatic incompatibility in *Didymium iridis* is regulated by two types of loci: fusion loci that mainly control cell mixing, and clear-zone loci that are involved in cytotoxic postfusion reactions. Carlile (**1972, 1973**), working with *Physarum polycephalum*, found the lethal effect to be modified by nutrition and plasmodial size; no lethal reactions occurred on plain agar, and only a few between microplasmodia. The nuclei of one strain nevertheless were eliminated. These heterogenic incompatibility mechanisms were interpreted as a recognition mechanism assuring exclusion of foreign nuclei (thus maintaining the genetic integrity of the plasmodium in nature) and a means of inhibiting the distribution of viruses and other harmful cytoplasmic components (Carlile, **1972**; Esser and Blaich, **1973**; Ling and Clark, **1981**). Similar mechanisms exist in many fungi and other eukaryotes (Esser and Blaich, **1973**).

When conditions which favor sporulation (see p. *18*) prevail, the plasmodium changes into one or more fruiting bodies characteristic of the species. This transformation is accompanied by cleavage of the protoplasm into uninucleate portions which become enveloped by walls and mature into spores. A single precleavage mitosis occurs, during which the nucleolus disappears and which (where known) is intranuclear, as in the plasmodium (Olive, **1975**). Capillitial development usually coincides with the beginning of sporulation.

In heterothallic species, each sporophore contains spores of both mating types. In addition, however, spores which behave as though they were homothallic are often present among the others.

The position of meiosis in the myxomycete life cycle has long been controversial, the evidence presented in the literature being contradictory. The two most prevalent views on the subject were: 1. Meiosis occurs in the developing sporangium preceding spore cleavage (Wilson and Cadman, **1928**; Schünemann, **1930**; Wilson and Ross, **1955**; Ross, **1961**b; Therrien, **1966**) and 2. Meiosis occurs in the spore soon after its formation, with all products of meiosis but one disintegrating, leaving the spore uninucleate and haploid (Stosch, **1935, 1937, 1965**; Aldrich, **1967**; Aldrich and Mims, **1970**; Aldrich and Carroll, **1971**). It is of course possible that meiosis does not occur in the same place in all species, but enough representative species have been investigated in this respect to give considerable assurance that when meiosis does occur it takes place in the young spores. Meiotic divisions are intranuclear and lack centrioles; the latter, together with Golgi apparatus appear in the spores after meiosis (Olive, **1975**). Interrupted or incomplete meiosis has been reported in *Fuligo septica* and postulated in *Stemonitis virginiensis* (Mims and Rogers, **1973**). In the former species, spores did not germinate until meiosis was completed much later, according to Aldrich and Blackwell (**1976**), and these authors suggested that a phenomenon of this type may explain the failure or difficulty experienced in germination trials for some strains or species, or account for the delay usually attributed to a necessary aging factor. The presence of an aberrant type of meiosis has recently been postulated even in haploid apogamic strains (Laane et al, **1976**).

The exogenous species. The life cycle of *Ceratiomyxa* differs in some important details from that of the endosporous species. Upon germination, the spore releases a quadrinucleate protoplast which soon elongates to form the "thread stage" (Gilbert, **1935**; McManus and McDade, **1961**; Nelson and Scheetz, **1975, 1976**), the significance of which is unknown. Eventually the thread rounds up and divides into four uninucleate segments which remain in close association and again divide forming an octette. The eight cells now become transformed into swarm cells. These fuse in pairs forming zygotes which develop into plasmodia. At the time of sporulation the plasmodium develops papillae from which the pillars, characteristic of *Ceratiomyxa fruticulosa*, are formed. A thin layer of protoplasm covers the fructification. After a nuclear division, the protoplasm cleaves into uninucleate segments, the protospores. The presence here of synaptonemal complexes (Furtado and Olive, **1971**), presumptive indicators of meiosis, suggests that reduction division probably begins at this time. The protospores are then elevated on threadlike stalks, become enveloped by a wall, and develop into spores. Meiosis is apparently completed in the young spores soon after their formation (Gilbert, **1935**).

Olive (**1975**) considered *Ceratiomyxa* more closely related to the protostelids than to the endosporous Myxomycetes and accordingly transferred the genus into his order Protostelida.

According to our present information, chromosome numbers in endosporous species of Myxomycetes appear to vary greatly, ranging from n = 4 to ca. 90 (Collins, **1979**). In *Ceratiomyxa fruticulosa*, the only exosporous species which has been investigated, the chromosome number is n = 8. In other species, polyploidy seems to be widespread (Collins, **1979**) and, under certain laboratory conditions and treatments, varying ploidy levels and apparently even aneuploidy are known to occur (Ross, **1966**; Koevenig and Jackson, **1966**; S. Kerr, **1968**; Mohberg, **1982**).

V. Physiology

Spore germination. The time required for spore germination as well as the percentage of germinating spores varies with the conditions, the age of the spores, the species, the strain, and even with the particular fruiting body. For most species investigated the optimum temperature for spore germination is 22°–30°C and the optimum pH, 4.5–7.0 (Smart, **1937**). Wetting and drying and extracts from various natural substrata have also been reported as favoring spore germination. Elliott (**1949**), working with 59 species and using (presumably herbarium) specimens of various ages, some of them as old as 61 years, induced the spores of all species but one to germinate by employing sodium taurocholate as a wetting agent. Erbisch (**1964**) reported "abundant" germination after 72 h in 75-year-old spores of *Hemitrichia clavata* treated with bile salt. Oxygen is a requirement for the germination of *Fuligo septica* spores (Nelson and Orlowski, **1981**), those being kept under anaerobic conditions failing to germinate until they were exposed to air.

Germination is accomplished by one of two methods. Either the spore cracks open or a minute pore dissolves on the wall and the protoplast emerges. The method of germination evidently is constant for each species. Lipid bodies appear to provide the energy source during germination (Alexopoulos and Mims, **1979**). The two methods are described in ultrastructural detail by Mims (**1971**) and Mims and Rogers (**1973**). Although germination of single spores was said to be more difficult than germination in mass sowings (Smart, **1937**), single-spore cultures of many species have been obtained, yielding, in heterothallic species, clones of myxamoebae or, in others, proceeding to complete the life cycle. In fact, Nelson and Orlowski (**1981**) found that, in *Fuligo septica*, dense concentrations of spores resulted in lower percentages of germination, suggesting the presence of an auto-inhibitor.

Flagellated cells and myxamoebae. In the normal course of events, both flagellated cells and myxamoebae are formed during the life cycle of a myxomycete. Environmental conditions seem to play an important role in determining the duration of each stage: free water induces the formation of flagella; dry conditions favor the myxamoebal stage. In artificial culture the flagellated stage may be completely suppressed, at least in some species, by germinating the spores on a moist agar surface in the absence of free water (Alexopoulos, **1960**a). By adding water to a culture or permitting a wet culture to dry, a shift from flagellated cells to myxamoebae and vice versa may sometimes be induced over a long period of time before zygote formation begins.

Cell division probably occurs only in the myxamoebal stage, the swarm cells withdrawing their flagella before dividing. The myxamoebal stage of *Didymium nigripes* may be prolonged and plasmodium formation prevented by the addition of 2% glucose or 0.2% brucine to the medium (Kerr and Sussman, **1958**), but zygote formation is not inhibited, at least by glucose (Therrien, **1966**). Zygote formation involves cell surface changes in the outer membranes, since zygotes *coalesce* with other zygotes, but *engulf* genetically identical but haploid amoebae (Ross, **1967**a).

The plasmodium. The rhythmic, reversible streaming of the protoplasm characteristic of the plasmodia of most Myxomycetes is a well-known phenomenon. An early theory seeking to explain protoplasmic streaming implicated the changes in viscosity of myxomyosin when it interacts with ATP (Kamiya, **1959**). Ultrastructural studies have refined this premise by clarifying the structures and mechanisms involved. Myxomyosin, actin, and ATP were demonstrated in the plasmodium of *Physarum polycephalum* (Ts'o et al, **1956**a, b; **1957**a, b) and the reaction of these substances appears to be similar to that of the actomyosin-ATP system in muscle. More recent experiments with the phaneroplasmodia of *Physarum polycephalum* have shown that protoplasmic streaming is caused by a hydraulic pressure flow mechanism generated by the contractions of the protoplasm (Komnick et al, **1973**). The latter are a function of the assembly and disassembly of actin-containing cytoplasmic fibrils in the ectoplasm (Hinssen, **1981**; Wohlfarth-Bottermann and others, cited by Hinssen). The differentiation into gel-like ectoplasm and fluid endoplasm depends on the degree of polymerization of the actin, which in turn was found to be regulated by a polymerization-inhibiting protein ("actin modulating protein"). What mechanism directs streaming in aphanoplasmodia and protoplasmodia apparently has not yet been determined.

Little is known about the cytoplasmic movement of myxamoebae, but

Taniguchi et al (**1978, 1980**) have isolated myosin also from that motile stage in *Physarum polycephalum*.

Streaming of the protoplasm in the plasmodium is directly related to locomotion. When the plasmodium is moving over the substratum, the total volume of protoplasm transported over a given period of time will be somewhat greater in the general direction of movement than in the reverse (Kamiya, **1950, 1959**). This is obviously a simple but effective method of circulation. Polarity of the plasmodium appears to be closely associated with potassium concentration, a greater concentration prevailing in the anterior over the posterior regions of a migrating plasmodium (Anderson, **1962, 1964**).

Turnock et al (**1981**) have found some 300 proteins contained in plasmodia and amoebae of *Physarum polycephalum*, of which ca. 3/4 were present in both stages, but synthesized in different proportions and at different rates.

The nature of the pigments in myxomycete plasmodia has attracted the attention of a number of students but, beyond the fact that many of these pigments act as indicators, changing color with changes of pH, little definite knowledge has been obtained. In nature, pigmentation is affected by a number of conditions (as indicated on p. 5), but under controlled laboratory conditions it appears to be a stable factor (Collins, **1979**). Collins and others, however, have isolated a number of color mutants in the laboratory. It has been both suggested and denied that the yellow pigments have the properties of anthracenes, flavones, pteridines, polypeptides, or polyenes. Czeczuga (**1980**) found carotenoids in the plasmodia of various species and xanthophylls in *Fuligo septica* plasmodia as well as in developing sporophores of other species. The pigments also have been postulated to be photoreceptors playing an important role in the process of sporulation (Lieth and Meyer, **1957**; Wolf, **1959**; Daniel and Rusch, **1962**a; Wormington and Weaver, **1976**).

The presence of various enzymes, vitamins, sterols, and other organic substances has been detected in the plasmodium of *Physarum polycephalum*, and the production of antibiotics by several species has been reported (Locquin, **1948**; Sobels, **1950**; Considine and Mallette, **1965**; Taylor and Mallette, **1978**; Chassain, **1980**). The responses of plasmodia to various external factors such as anaesthetics, low and high temperatures, gravity, light, and irradiation have been studied to some extent and considerable knowledge has accumulated on this subject. Rakoczy (**1973**), working with *Physarum nudum*, found that light affected direction of migration, pigment composition, and sporulation, and inhibited plasmodial growth. Furthermore, plasmodial age influenced direction of migration (with respect to light) and the length of exposure to light required to induce sporulation. Chemotaxis (Carlile, **1970**; Madelin et al, **1975**; and earlier reports cited by these authors), nutritional state of the plasmodium, and humidity also control migrations (Hüttermann, **1973**a).

Sporulation. In most species, sporulation in nature appears to occur largely at night (Gray and Alexopoulos, **1968**; Dee, **1975**; and several earlier workers). Temperature, moisture, availability of food, light, plasmodial size, and pH are all factors which influence sporulation, but the initial stimulus which induces this process is still unknown. Gray (**1939**), working with *Physarum polycephalum*, showed that temperature and pH were interrelated factors. Within certain limits, the higher the temperature, the lower the pH required for sporulation. Gray (**1938, 1941, 1953**) also first showed that light is necessary for the sporulation of *Physarum polycephalum* plasmodia. Starvation is necessary to bring about sporulation in some species (Dee, **1975**; Olive, **1975**), but apparently not all. Ultrastructural studies on certain species in family Trichiaceae, for example, revealed viable bacteria from the culture medium within food vacuoles contained in the protoplasm of young sporophores (Mims, **1969**), the slime coat surrounding the plasmodiocarp (Charvat et al, **1973**), and within the capillitium (Charvat et al, **1974**). Age appears to be a factor (Olive, **1975**). Daniel and Rusch (**1962**a, b), working with bacterium-free cultures of *Physarum polycephalum* found that the conditions necessary for sporulation are: 1. A sporulation medium containing niacin and niacinamide or certain substitutes such as tryptophane; 2. An optimal growth age; 3. A dark incubation period of four days; and 4. A subsequent exposure to light of wavelengths between 350 and 500 mμ. Sporulation-competent plasmodia may also be obtained without illumination, by injection of cytoplasm from another plasmodium which was activated by light, or by small amounts of salt solutions (Wormington et al, **1975**, quoted by Gorman and Wilkins, **1980**). Starvation in darkness leads to spherule formation (Sauer et al, **1969**). These authors enumerate the synthesis of DNA *before* illumination, a *continued* synthesis of proteins, and RNA synthesis until 3 h after the end of illumination as essential processes in sporulation. At that time the plasmodium (of *Physarum polycephalum*) becomes irreversibly committed to sporulation. A mitosis also appears to be prerequisite to this transformation.

Investigations continue on the biochemical changes which occur during sporulation and a large body of information is available. A shift in oxidases apparently takes place. A higher cytochrome-oxidase activity occurs in the spores than in the plasmodium and a greater ascorbic acid oxidase activity in the plasmodium than in the spores (Ward, **1958**). Gorman and Wilkins (**1980**) point out certain similarities between spherulation (sclerotium formation) and sporulation: change from glycogen synthesis to utilization, changes in protein metabolism, loss of DNA, RNA, and nuclei, and calcium deposition (in *Physarum*; what happens in the nonlimy species is not yet known). In addition, while the spore walls include melanin, spherules contain melanin precursors. These comparisons led the cited authors to regard spherulation "in several respects, biochemically and cytologically to be an incomplete form of sporulation." Sporulation differs from spherulation by being an irreversible transformation and by requiring light for induction (in *Physarum*). For updated detailed treatments of these topics, as well as cytology, genetics, and physiology consult Dee (**1975**), Olive (**1975**), Collins (**1979**), Dove and Rusch (**1980**), and Aldrich and Daniel (**1982**).

VI. Laboratory Culture and Nutrition

Plasmodia of Myxomycetes may be brought into the laboratory from the field and may often be induced to spread and grow on artificial media. Plasmodia may be maintained for a long time in culture by feeding them oat flakes (Camp, **1937**). The difficulties encountered in growing Myxomycetes in the laboratory starting with spores, however, have plagued experimentalists for a long time. Nevertheless, much progress has been made in recent years and about 75 species — most of them in the order Physarales — have now been induced to complete their life cycles on artificial media in crude culture. Monoxenic cultures of many of these organisms may be easily established by spreading spores on an agar medium containing penicillin and streptomycin (Gray and Alexopoulos, **1968**), allowing the myxamoebae issuing from the spores to migrate away from contaminating organisms, and transferring them to a suitable medium together with a suspension of a known bacterium such as *Escherichia coli* or *Enterobacter aerogenes*. Such cultures proceed to develop plasmodia which eventually sporulate. Corn meal agar, Knop's solution agar, or lactose-yeast extract agar give good results. The agar medium should be suitable for bacterial growth but not so rich as to permit the bacteria to overwhelm the slime mold. Recent research indicates that a medium buffered at the proper pH can improve plasmodium production and sporulation (Collins and Tang, **1973**; Mock and Kowalski, **1976**). A helpful guide for anyone interested in attempting laboratory culture and temporary or permanent mounts of the various stages is Collins' (**1973**) treatment in the *Encyclopedia of Microscopy and Microtechnique*.

Axenic cultures are more difficult to establish. Although a few of the early workers claimed to have grown myxomycete plasmodia in pure culture, it was Cohen (**1939**) who first set up rigid standards for testing purity and who obtained undoubted axenic cultures of several species. Subsequently, several workers succeeded in purifying plasmodia by frequent transfers, use of antibiotics, etc., but few species may be maintained in bacterium-free cultures for a long time. The success of Rusch and his coworkers (Daniel et al, **1962**) in growing the plasmodium of *Physarum polycephalum* in bacterium-free culture, in liquid media of known chemical composition, has made possible exact nutritional studies as well as studies on conditions necessary for sporulation. Some years later Henney and Lynch (**1969**) grew the plasmodia of *Physarum flavicomum* and *P. rigidum* in axenic culture in chemically defined liquid media.

Relatively recently we learned how to grow myxamoebae in axenic culture (Ross, **1964**; Henney and Henney, **1968**; Haskins, **1970**; Goodman, **1972**). It is hoped that this will eventually lead to the axenic culture of the entire life cycle of one or more species on the same medium. Ultimate success will be achieved when surface-sterilized spores can be sown on a chemically defined medium and induced to develop the entire life cycle as is done with many fungi. It should be kept in mind, however, as cautioned by Ross (**1979**), that axenic culture may lead to a loss of certain natural or wild-type characteristics. For example, axenically maintained amoebae may, after some years, develop many ploidy levels and lose the ability to form plasmodia (Ross, **1966**).

Much progress has been made since 1968 in the study of the biochemistry of these organisms and a large number of papers have been published. A review of biochemical studies is outside the scope of a taxonomic treatise such as this, however.

VII.
Ecology and Geographic Distribution

Most species of Myxomycetes are cosmopolitan. Moisture and temperature appear to be the chief factors governing the abundance of Myxomycetes in any particular region. No known species are strictly aquatic or strictly xerophilic, but collections of Myxomycetes have been made at certain times from bogs, from a brook (Gottsberger and Nannenga-Bremekamp, **1971**), and from desert areas. Some appear to be confined to the tropics or subtropics and a few to the Temperate Zone. *Physarum nicaraguense* and *P. javanicum*, for example, have been found only in warm climates. *Hemitrichia clavata*, on the other hand, appears to be limited to the temperate regions; although it has been reported from the tropics, careful examination of all such specimens has proved them to be *H. calyculata*, a related but distinct species. A few slime molds such as *Diderma niveum*, *Lepidoderma granuliferum*, and *Lamproderma carestiae* appear to be strictly alpine or subalpine in their distribution.

Many species are seasonal in appearance of their fructifications. In the Temperate Zone, some sporulate early in the spring and cease by the middle of the summer; others begin in the summer and continue until fall (Krzemieniewska, **1957**; Mitchel et al, **1980**). Whether this is a question of photoperiodism or a response to temperature, moisture, or some other factors is not known.

Although most species appear to be independent of the substratum on which they sporulate, some show a distinct preference for one type of substratum or another. Some Badhamias, for example, most often fructify on bark of deciduous trees; some Cribrarias are partial to coniferous wood; many Didymia sporulate mostly on dead leaves; most Trichias on dead wood. A small number of species appear to be exclusively or predominantly fimicolous (Eliasson and Lundqvist, **1979**). None of these correlations are absolute, but they occur too often to be entirely coincidental.

Whereas plasmodia and fructifications of Myxomycetes are particularly abundant in moist, forested areas (at least in temperate regions), the typically wind-disseminated spores are present almost everywhere. In addition to air currents, arthropods have been linked with spore dispersal of certain species (Keller and Smith, **1978**; Blackwell et al, **1982**).

By employing special laboratory techniques, slime mold plasmodia or sporophores may be developed from a great variety of materials, ranging from rain water (Pettersson, **1940**) to desert debris (Evenson, **1962**; Blackwell and Gilbertson, **1980**), representing a diverse range of conditions. Many species may also be easily cultured by exposing to the wind agar plates or coated slides from which isolations may be made (Brown et al, **1964**).

The moist chamber technique, first introduced for bark cultures by Gilbert and Martin (**1933**) and used since then successfully by many investigators, has expanded vastly our knowledge of the geographic distribution of many species, particularly those possessing minute sporangia and hence generally overlooked in the field. By the use of this technique, some species hitherto considered to be rare have been shown to be common and widely distributed.

VIII.
Collection and Care of Specimens

Throughout the world, specimens may be gathered at the proper season in almost any locality. After the last spring frost, plasmodia are to be found everywhere on piles of organic refuse: in the woods, especially about fallen rotting logs, undisturbed piles of leaves, beds of moss, stumps, by the seeping edge of melting snow on mountain sides, by sedgy drain or swamp, or in the open field and garden where piles of straw or herbaceous matter of any kind are decaying. In any locality the plasmodia pass rapidly to fructification, but not infrequently a plasmodium in June will be succeeded in the same place by others of the same species, until the cold of approaching winter checks all vital phenomena. The process of sporulation should be watched as far as possible and, for herbarium material, allowed to pass to perfection in the field.

Many species, such as *Licea biforis*, *L. kleistobolus*, *Echinostelium minutum*, *Comatricha fimbriata*, etc., often overlooked in the field because of the minute size of their sporangia, may be spotted by carrying a dissecting microscope on collecting trips or, as mentioned earlier, they may be developed in moist chamber culture on bark taken from living trees, dead wood, decaying leaves, seed pods, conifer needles, other plant debris, and animal dung. The procedure is simple. Petri dishes or small covered bowls are fitted with 9 cm filter paper discs (or paper towels cut to size) and sterilized in the oven or autoclave. The material to be cultured is placed on the filter paper, covered with sterile distilled water, and allowed to soak overnight. The water is then poured off and the cultures are incubated for a week, ten days, or considerably longer. Water may be added as needed to keep the chambers moist. Plasmodia and fruiting bodies develop in a remarkably large percentage of such cultures and may be detected under a good stereomicroscope. It should be pointed out here that mature sporangia of some species, e.g. *Echinostelium elachiston*, sometimes become evident within 24 h from the time the material is placed in moist chamber. Unless such sporangia are properly dried they eventually turn moldy, disintegrate, or are destroyed by mites or other minute animals. It is important, therefore, to begin examining the cultures the day after the material is wet.

Specimens collected in the field should be placed immediately in boxes in such a way as to suffer no injury in transport; beautiful material is often ruined by lack of care on the part of the collector. Once at the herbarium, specimens may be mounted by affixing the supporting material to the inside of the cover of a small box so that the label and the specimen will be inseparable. White glue is usually used. Sporophores developed in moist chamber culture should be dried very gradually after they have fully matured, before mounting for storage. Boxes of uniform size and depth are most desirable. In the National Fungus Collections and some other herbaria, specimens are mounted in boxes 4½ × 10½ × 2 cm (1¾ × 4 × ⅞ ins) or, for larger material, 10 × 9¼ × 2 cm. The shallow cover, inside which the specimen is mounted, permits ready examination of the material with lens or stereomicroscope. Such boxes fit snugly in a shallow, covered tray the size of an herbarium sheet, and five trays fit easily into a standard herbarium shelf.

The method, while somewhat wasteful of space in the case of very small collections, is admirably suited to the great majority of collections and permits their filing in strict order. Every care must be taken to exclude insects. A small box or glass container with paradichlorobenzene crystals (PDB) placed at the top shelf of each herbarium case is an efficient remedy. There is, however, some good evidence that the use of this fumigant delays or totally prevents germination of the spores. Material which is intended for culture work should, therefore, not be stored in the presence of PDB.

Lyophilization deserves mention as a promising means of preservation and laboratory storage of the myxamoebal state (Davis, **1965**; N. Kerr, **1965**; Collins, **1979**).

For simple microscopic examination it will be found convenient first to wet the material with absolute alcohol; then, before the ethanol has completely evaporated, to add a 2–3% solution of KOH to cause the spores and other structures to assume plumpness. The KOH solution should not be stronger than this or the material will swell too much. This solution should be blotted off, using bits of filter paper to remove the excess, but not so as to permit the material to dry. A drop of 8% glycerine may then be added and a cover glass mounted. For permanent slides several methods have proved useful. In using the glycerine method just described, the water in the glycerine may be allowed to evaporate either before or after the cover slip is added and the mount then sealed. Keeping the slides free from dust while the water is evaporating is important; placing them in petri dishes is convenient. Every effort should be made to eliminate air bubbles since, when these are present, they tend to move to the vicinity of the material in the mount. When the glycerine has become concentrated, all excess should be removed and the cover slip may be ringed with a suitable cement. A modification of this, giving slides which will resist handling even better, is to make the mount on a 22 mm cover slip and when the glycerine has reached proper consistency, cover it with a 15 mm circle, then invert the mount on a standard slide with regular mounting medium (Diehl, **1929**). Such slides remain usable for many years. Glycerine jelly is even better than glycerine, but since it must be used warm it is less convenient. Many prefer Hantsch's fluid or Amann's medium[1] which are handled in the same way except that the slide should be heated gently. Another mounting medium which has been used with some success is Hoyer's.[2] Specimens may be mounted directly in this medium or may first be wet with 95% ethanol to eliminate air bubbles. Hoyer's medium preparations are semipermanent. After two or three years they dry out, making it necessary to remount the specimens. Noncalcareous species may be mounted directly in Turtox CMC-10 or

---

1. Hantsch's fluid

Ethanol 90%	3 parts
Water	2 parts
Glycerine	1 part

Amann's medium

Phenol	20 gms
Lactic acid	20 gms
Glycerine	40 ml
Water	20 ml

2. Hoyer's medium

Distilled water	50 ml
Arabic gum lump	30 gms
Chloral hydrate	200 gms
Glycerine	20 gms

Soak Arabic gum for 24 hours. Add chloral hydrate and let solution stand until all material dissolves. This may require several days before the glycerine is added and the solution is ready for use.

CMC-S mounting medium and heated gently. Such preparations dry hard and last for a long time.

Unfortunately, none of these materials can be counted upon to preserve calcareous elements although in some species they keep fairly well. Lime nodes may be preserved by mounting in a solution consisting of 100 ml white Karo syrup, 300 ml water, and 40% formalin (Stevens, **1974**, p. 652). However, in specimens mounted in any but the glycerine media, the spores tend to shrink. An ideal mounting medium has yet to be found. All mounts should be sealed, either with nail polish or some more permanent sealer such as Canada balsam or Zut, the latter a product of Bennett Paint Co., Salt Lake City, Utah.

IX. Taxonomy

There has been no general agreement in the past as to the limits of the class Myxomycetes. A number of groups of uncertain position have been regarded as related to the true slime molds by some authors, but excluded by others, Only thirty years ago, Bessey (**1950**) grouped the Myxomycetes, using for them the ordinal name Myxogastrales, together with the Acrasiales, the Plasmodiophorales, and the Labyrinthulales, in the subclass Mycetozoa, class Sarcodina, phylum Protozoa. All four of these groups have received considerable attention in the past quarter century and much information concerning their morphology, life history, and physiology has accumulated through observation and experiment. Such knowledge has widened rather than narrowed the gaps among some of these organisms and few biologists who have studied any one or more of these groups now seriously consider them to be closely related. It is unfortunate, therefore, that the term slime molds has been used for all these groups, but it is so well established in the literature that modifying adjectives have become necessary for clarification. Thus the Acrasiales are the cellular slime molds; the Plasmodiophorales, the endoparasitic slime molds; and the Labyrinthulales, the net slime molds. The Myxomycetes have been variously referred to as acellular, noncellular, plasmodial, or true slime molds. Since the discovery of the protostelids, some of which have simple plasmodia, only the term "true slime molds" remains as exclusively characterizing the Myxomycetes.

The Acrasiales, discovered by Brefeld (**1869**) but first studied as a group by Tieghem (**1880**), resemble the Myxomycetes in the possession of a naked amoeboid stage, but produce neither swarm cells nor true plasmodia. The amoebae become aggregated but do not lose their identity; may, in fact, readily be shaken apart in water. Bonner (**1967**) and, more recently, Olive (**1975**) have summarized our knowledge of these organisms.

The Plasmodiophorales include several genera, all parasitic on vascular plants or fungi. They show several points of resemblance to the Myxomycetes, especially in the character of the heterokont swarm cell and its apical, whiplash flagella, and in the assimilative stage, which is a true plasmodium. On the other hand, they entirely lack the secreted sporangial wall and the capillitial threads found in most of the true slime molds, they form zoosporangia comparable to those of many "Phycomycetes," they have a strictly endoparasitic life style, and they are held to possess a distinctive type of nuclear division. Schroeter (**1886**, p. 133) erected the order Phytomyxini to contain them and his classification has been extensively followed. The tendency among more recent authors, e.g. Fitzpatrick (**1930**), Gäumann (**1926, 1949, 1964**), Sparrow (**1943, 1960**), and Alexopoulos (**1962**) has been to include them among the lower "Phycomycetes," to place them in a class by themselves, the Plasmodiophoromycetes (Sparrow, **1959**; Webster, **1970**; Alexopoulos and Mims, **1979**), or in a separate subphylum Plasmodiophorina in phylum Gymnomyxa of the kingdom Protista (Olive, **1975**; Ross, **1979**). Karling (**1968**) and Olive (**1975**) have updated our knowledge about these organisms.

The order Labyrinthulales is a curious group of organisms of uncertain position. The assimilative stage consists of spindle-shaped or ovoid cells gliding on or within slime tracks or tubes which are united to form a network known as the filoplasmodium or net plasmodium; hence the name net slime molds, often given these organisms. The filoplasmodium is in no way comparable to the plasmodium of the Myxomycetes. Olive (**1975**) refers to this stage as "ectoplasmic net" and on the basis of ultrastructural evidence believes that the labyrinthulas are closely related to the thraustochytrids. *Labyrinthula minuta*, however, is said to produce multinucleate protoplasts which may be true plasmodia (Watson and Raper, **1957**).

In 1932, Martin presented his views regarding the fungal relationships of the Myxomycetes and, in his 1949 treatment of the Myxomycetes in North American Flora, placed this class in the division Fungi. Later (**1961**a) he reiterated his position regarding the phylogeny of the group and strengthened his views by citing new evidence which had been obtained in the interim. Martin and Alexopoulos (**1969**), Alexopoulos (**1973**), and others perpetuated this placement.

Considerable recently obtained evidence points to a close relationship of the Myxomycetes to the protostelids, a group described by Olive (**1964, 1967**) and Olive and Stoianovitch (**1966**), as order Protostelida in the Mycetozoa. This viewpoint is reflected in both Olive's (**1975**) and Alexopoulos and Mims' (**1979**) recent classification schemes. The Protostelida exhibit some of the characters of the Myxomycetes but cannot at the moment be included in this group. All produce extremely minute fruiting bodies with one or two spores. *Cavostelium* possesses amoebae which become converted into anteriorly flagellated cells. It produces mostly uni- or bispored sporophores but forms no plasmodia. The other two genera produce spores, amoebae, and plasmodia, but no flagellated cells. In *Schizoplasmodium* the spores behave as ballistospores. Gametic fusion has not been observed in any of the Protostelida. Olive and Stoianovitch (**1971**) and Hung and Olive (**1972**) found ultrastructural similarities in *Cavostelium* and the myxomycete genus *Echinostelium*. On the basis of his studies, Olive (**1975**) placed all five orders of Myxomycetes in one straight evolutionary line from the flagellate Protostelia via *Cavostelium* of the latter group through *Echinostelium*. From this genus he envisioned a succession to the more complex stalked forms, then the sessile types of sporophores.

Certain members of the Myxomycetes had been noted by careful observers for over three hundred years. Lister (**1925**) cited Pankow's figure and description, 1654, of the species now known as *Lycogala epidendrum* (L.) Fr. Ray, in 1660, called the same species *Fungus coccineus*, etc.; Ruppius, in 1718, *Lycoperdon sanguineum*, etc.; Dillenius, a year later, *Bovista miniata*; Buxbaum, in 1721, *Lycoperdon epidendron*. In 1729, Micheli erected the genus *Lycogala* for it and at the same time added recognizable descriptions and illustrations of several other genera and even species. But Micheli's light was too strong for his generation. As Fries (**1829**, p. 73), writing a century later, said "Immortalis Micheli tam claram lucem accendit, ut successores proximi eam ne ferre quidem potuerint." Notwithstanding Micheli's clear distinctions, he was entirely disregarded and *Lycogala* was dubbed *Lycoperdon* and *Mucor* down to the end of the century. It was not until 1794 that Persoon came around to the standpoint of Micheli and wrote *Lycogala "miniata,"* although Adanson had adopted the generic name in a somewhat uncertain application in 1763. Fries himself, reviewing the labors of his predecessors, grouped the slime molds as a suborder of the Gasteromycetes, although clearly recognizing the peculiar character of their assimilative phase, and gave expression to his view of their nature and position when he named the suborder Myxogastres. In 1833, Link, perceiving more clearly the distinctness of the group, substituted the name Myxomycetes. Wallroth (**1833**) used the name in the same year and he is usually credited with it, but he seems strangely to have confused its limitations, apparently regarding it as a synonym for the Gasteromycetes of Fries. Link's usage passed unchallenged for over a quarter of a century. The slime molds were apart by themselves; they were fungi without question and, of course, plants.

In 1858, A. de Bary published the first of his noteworthy studies on the Myxomycetes, based upon careful observation of their life cycles and particularly upon the transition between the plasmodial and sporulating stages. These studies were greatly amplified in 1859 and 1864. As a result of his investigations Bary concluded that the relationships of the slime molds were with the amoeboid Protozoa rather than with the fungi, and to emphasize his viewpoint, he proposed the name *Mycetozoa* — fungous animals.

In 1884, he modified the group so as to include not only the Myxomycetes of Wallroth, but also the Acrasiées of Tieghem. Bary's name for the group has, with varying limitations, since been adopted by many distinguished authorities, including Rostafinski, Saville Kent, Zopf, the Listers, Lankester, Hagelstein, and Olive.

More recently Copeland (**1956**), recognizing the kingdom Protoctista, divided it into eight phyla of which the Protoplasta includes five classes, among them the Mycetozoa and the Sarkodina. The former consisted of three orders, two of which were made up of the Myxomycetes and the third of which encompassed the Plasmodiophorales designated by Copeland as Phytomyxida. In the latter class Copeland placed the Labyrinthulales as the family Labyrinthulida, and the Acrasiales as the family Guttulinacea, together with five families of amoebae, in the order Nuda.

A more recent general taxonomic treatment which included the slime molds was that proposed by Jahn and Bovee (**1965**). These authors divided the Sarcodina into two classes on the basis of the mechanism of movement: the Autotractea "with slender, filamentous pseudopods in which two-way streaming of cytoplasm is detectable" and Hydraulea "with tubular or polytubular body and pseudopods in which the gel tube contracts to drive the more fluid inner contents." The cellular slime molds were placed in the former, in the order Acrasida, and the true slime molds (Myxomycetes) were located in the latter, in the order Mycetozoida. The Labyrinthulales and the Plasmodiophorales were not mentioned. Finally, Whittaker (**1969**), using a five-kingdom system, placed the Myxomycetes in the phylum Myxomycota alongside the phyla Acrasiomycota and Labyrinthulomycota in the subkingdom Gymnomycota of his kingdom Fungi. Ainsworth (**1973**) listed the divisions Myxomycota and Eumycota of the kingdom or subkingdom Fungi. Olive (**1975**) recognized the two kingdoms Fungi and Protista, placing the "mycetozoans" as subphylum Mycetozoa in the phylum Gymnomyxa. The subphylum Mycetozoa consisted of the classes Acrasea and Eumycetozoa, with the latter divided into subclasses Protostelia, Dicytostelia, and Myxogastria, the last one comprising the true slime molds. Alexopoulos and Mims (**1979**) accepted the kingdom Fungi and recognized three divisions within the kingdom. The first of these, Gymnomycota, contained two subdivisions, the first (Acrasiogymnomycotina) including the two classes Myxomycetes and Protosteliomycetes. Olive (**1975**), on the other hand, stressed the predominantly phagotropic mode of nutrition, the absence of a cell wall in the assimilative phase, the nature of the flagellar apparatus, and the difference in basic nuclear proteins, as evidence against a close relationship of the true slime molds with the fungi. In the absence of fossil records, any viewpoint on the phylogeny of the Myxomycetes is obviously highly speculative. Diagrams for possible evolutionary schemes have been constructed by Alexopoulos (**1969, 1973, 1982**), Olive (**1975**), and Collins (**1979**).

Whatever the position of the true slime molds among living organisms may be, their actual study had been left almost entirely to the botanists, and particularly the mycologists, until recent years when biochemists, biophysicists, and geneticists recognized the importance of the Myxomycetes as tools in the study of fundamental biological processes, such as nuclear division, DNA synthesis, morphogenesis, protoplasmic streaming, cell fusion, etc. The taxonomy and nomenclature of the group continue, however, to remain within the domain of the mycologists. By vote of the International Botanical Congress of Vienna (**1905**), accepted by all subsequent congresses, the nomenclature of the group is fixed as beginning with Linnaeus' *Species plantarum* of 1753. Linnaeus, to be sure, knew little about the fungi or slime molds, and apparently cared less. Nevertheless, the fixing of this date permits taking into account the work of a number of active students of this group dating from the closing years of the 18th century. Chief among these is perhaps Bulliard, in whose extensive work "Histoire de Champignons de la France" (**1791**) may be found a number of recognizable descriptions and illustrations of slime molds, unexcelled up to that time. Noteworthy references to certain species were published still earlier between 1753 and 1791 by Gleditsch (**1753**), Schaeffer (**1762–1774**), Müller (**1777**), Batsch (**1783–1789**), Leers (**1789**), and others. Since that time a host of students has given more or less attention to the group, of whom the outstanding names up to the time of Rostafinski are Hoffman, Schrader, Sowerby, Persoon, Fries, Ehrenberg, Link, Fuckel, Schweinitz, Berkeley, Curtis.

The greatest taxonomic advance after Fries is embodied in the monographic treatment of Rostafinski, whose "Versuch" of 1873 was followed by the monograph of 1874–1875 and its supplement of 1876. Rostafinski, a pupil of Bary, followed the latter's example of making intensive use of the microscope, which by that time, of course, was greatly improved over the crude instruments at the command of earlier workers, although optically much inferior to modern apochromatic and immersion lenses. The monograph and supplement, written in Polish, were largely inaccessible to students in other countries, but were made available to English-speaking researchers to a considerable extent by the works of Cooke (**1877**) and Berlese (**1888**). In 1892, Massee published his monograph, based on that of Rostafinski but departing in many particulars from the latter's treatment, and greatly increasing the number of recognized species, not infrequently on an insufficient basis. Two years later appeared the first edition of the standard English monograph, A. Lister's "Mycetozoa," revised in 1911 and again in 1925 by his daughter, G. Lister. The illustrations in this work, many of them in natural colors in the later editions, have not been surpassed in comprehensiveness and general accuracy, and it is not surprising that European treatments in other languages have largely been modelled upon this excellent volume. Most noteworthy of these is Schinz's (**1912–1920**) in Rabenhorst's Kryptogamen-Flora.

In North America the first extensive collections and reports were made by Schweinitz (**1822, 1832**). Later active collectors were Curtis, Ravenel, Ellis, Peck, Farlow, Morgan, Rex, Wingate, Thaxter, Bethel, Sturgis, and Bilgram. Cooke, in 1877, published the first general account of the slime molds of the United States, followed by that of Morgan (**1893–1900**). The accounts of the slime molds of eastern Iowa (**1892, 1893**a) and of Nicaragua (**1893**b) by Macbride preceded the first edition of his *North American Slime-moulds* (**1899**). The greatly enlarged and emended second edition of this work (**1922**) became the basis for *The Myxomycetes* (**1934**) by Macbride and Martin. The new classification of the endosporous species into four orders, which Macbride had been developing, then became well established and was accepted until the nineteen-sixties by the majority of mycologists in the Western Hemisphere. In 1944, Hagelstein published his *Mycetozoa of North America*, based on the great collection of slime molds he, together with Rispaud and others, had amassed in the New York Botanical Garden. Hagelstein adopted Lister's classification and Lister's keys, from which Macbride and Martin had deviated. Volume I, Part 1, of *North American Flora*, dealing with the Myxomycetes, was published in 1949. In this treatise Martin followed Macbride and Martin's classification for the most part, but changed the order of treatment. The world monograph on these organisms by Martin and Alexopoulos (**1969**) is based on Macbride and Martin's monograph and on Martin's treatment in *North American Flora*. This system was modified by Alexopoulos (**1973**) to incorporate the third subclass, Stemonitomycetidae. The present abridged treatment encompasses the newer classification and additional information that has accumulated since 1969.

In the meantime, collectors in many countries have been active in the study of the Myxomycetes, and many regional treatises have been published. The most noteworthy of these are Thind's (**1977**) and Lakhanpal and Mukerji's (**1981**) treatments of Indian slime molds; Emoto's *Myxomycetes of Japan* (**1977**), whose excellent color plates beautifully illustrating 250 species are a fine addition to the iconography of the group; Kremieniewska's *Sluzowce Polski* (**1960**); the treatise on the Myxomycetes of Denmark by Bjørnekaer and Klinge (**1963**); *De Nederlandse Myxomyceten* by Nannenga-Bremekamp (**1974, 1979**); Arambarri's publication on Myxomycetes of Tierra del Fuego (**1975**); and Farr's monograph on neotropical Myxomycetes (**1976**b).

In the matter of nomenclature, the attempt has been made to adhere to the rules of the International Code of Botanical Nomenclature, in accordance with the philosophy expressed earlier (Martin and Alexopoulos, **1969**). According to the Code, Art. 10 and 16 (Stafleu, **1978**), priority in nomenclature is not mandatory for taxa above the rank of family. No attempt is made, therefore, to list all names which may be regarded as approximately or even strictly synonymous with orders or higher categories here recognized. An excellent list may be found in Copeland (**1956**), pp. *171–177*. Furthermore, the names applied to groups and the relative rank assigned to them has involved terminological alterations which represent in many cases little or nothing more than orthographical changes designed to bring the names into accord with contemporary usage. It seems advisable to keep such citations to a minimum.

Even at family levels, the problem is not simple. Too many families have been proposed differing markedly in basis, range, and circumscription, and they may well be radically revised in the future. Under such circumstances it has seemed desirable to cite names which reflect current treatments, beginning chiefly with the first edition of the Lister monograph (**1894**), and not to attempt an exhaustive series of citations at that level.

X. Use of the Keys

Despite the recent advancement in knowledge of the plasmodial stage, it has not yet proved possible to use plasmodial characters extensively in classification. It is still necessary to base this almost entirely upon the characters of the sporulating stage, which is relatively constant, most commonly observed, and may be preserved indefinitely. However, allowance must be made for the extraordinary sensitivity of these organisms to environmental conditions during the relatively short time in which the fructifications form, and which may be reflected in size, shape, capillitial development, amount and nature of lime deposited, spore size and intensity of markings, and practically every other factor which is used in the keys. Even mature fructifications in the field may be subjected to environmental conditions, notably wetting and drying, which may modify them in various ways. Despite these difficulties, most species present a fairly distinctive even though rather variable picture of what we are entitled to regard as properly referable to that category.

The distinctions between aethalium, pseudoaethalium, sporangium, and plasmodiocarp are not and cannot be made sharp. Nearly all recognized species have a morphological expression which is the commonest and therefore the most characteristic form of the species concerned; numerous species which are extensively collected show wide variations in form connected by complete series of intermediates. To say that the common form is "normal" and the deviations "abnormal" is to oversimplify the situation. It is not uncommon in large fruitings which appear to have developed from a single plasmodium to find, on the lower side of a log, sporangia which are provided with well-developed stalks, while those higher up have shorter stalks and those on top are sessile or plasmodiocarpous, all completely intergrading.

The obvious inference is that the sporophores developed in the more sheltered situation have had more favorable conditions and that the others matured under suboptimal conditions. But the spores of all may be equally viable and the plasmodia which develop from them have all the potentialities of the parent plasmodium. The increasing use of cultural methods in recent years has emphasized this clearly (Alexopoulos, **1969**).

Many species are strikingly colored, particularly when freshly formed, but even from the beginning there is great variation in this respect. Moreover, such colors may fade or become dingy with age and then be difficult to distinguish. Furthermore, color perception and interpretation vary to some extent among individual observers. It is for these reasons that both in the keys and in descriptions generalized color names are preferred to the more specialized terms employed in color keys.

The size, color, and markings of the spores may also reflect to some degree the influence of the environment during their maturation, although in well-matured specimens this is on the whole less evident than variation in the other features mentioned. It is, of course, a common experience to find large sporelike bodies amongst the spores. These are often quite obviously the result of overhasty drying during the maturation period and are frequently accompanied by other peculiarities suggesting the same thing. (Such sporelike bodies should not be confused with the distinctive vesicles consistently present in several species). But even where the spores of a specific collection are all alike they may differ in noticeable ways such as size range, depth of color, and surface markings, from those of another collection of the same species, which may have matured under different conditions. This is well brought out by examining a large series of specimens of any common and widespread species such as *Ceratiomyxa fruticulosa*, *Fuligo septica*, or *Arcyria cinerea*. Aldrich and Blackwell (**1976**) found high degrees of variation in their SEM comparisons of specimens and even of samples within individual collections. They therefore recommend extreme caution in designing taxonomic studies utilizing ultrastructural methods and characters. The numerous varietal names that clutter the literature are, in the great majority of cases, no more than names given to various expressions of characters which merge completely into those of the so-called "typical" forms. It is, of course, very difficult to express these variations in words which will at the same time describe a species clearly, and these difficulties quite naturally extend to the construction of dichotomous keys.

XI. Myxomycete Herbaria

Following is a list of institutional herbaria housing the most important myxomycete collections of the world. (This does not necessarily imply availability of the collections for study. Some institutions will send loans only to other institutions, some will not lend at all.) The institutions are listed alphabetically by official acronym (Holmgren and Keuken, **1974**). In addition to the institutional herbaria listed, many well-known authorities and collectors maintain private collections, not included here.

BM, British Museum (Natural History), Cromwell Road, London SW 7 58D, Great Britain. Lister collections, Linnaean types, Ravenel collections.

BO, Herbarium Bogoriense, Jalan Raya Juanda 22–24, Bogor, Indonesia. Boedijn collections.

BPI, National Fungus Collections, Mycology Laboratory, Bg. 011A, B.A.R.C.-West, Beltsville, MD 20705. Collections of Macbride, Martin, Morgan, Rex, Wingate, Bethel, Brooks, Alexopoulos, etc.; varying amounts of unicates, duplicates, and exsiccati from nearly all early and many contemporary major collectors.

C and CP, Botanical Museum and Herbarium, Gothersgade 130, DK-1123, Copenhagen K, and Department of Plant Pathology, Thorwaldsensvej 40, Entrance 8, DK-1871, Copenhagen V, respectively. Raunkiaer, Rostrup, Bjørnekaer, and other Danish collections.

CUP, Plant Pathology Herbarium, Cornell University, Ithaca, NY 14850. Muenscher collection.

DBG, Denver Botanic Gardens, Herbarium of Fungi, 909 York St., Denver, CO 80206. Colorado Myxomycetes (collections of D. H. Mitchel and S. W. Chapman).

DUH, Mycological Herbarium, Department of Botany, University of Delhi, Delhi — 110007, India. Lakhanpal and Mukerji collections.

FH, Farlow Library and Herbarium of Cryptogamic Botany, Harvard University, 20 Divinity Ave., Cambridge, MA 02138. Collections of Curtis, Patouillard, and others.

K, The Herbarium and Library, Royal Botanic Gardens, Kew, Richmond, Surrey TW9 3AE, Great Britain. Collections of M. C. Cooke, Berkeley, and others.

L, Rijksherbarium, Schelpenkade 6, Leiden, Netherlands. Persoon collections.

LAU, Musée botanique cantonal, Palais de Rumine, Lausanne, Switzerland. Meylan collections.

LINN, The Linnaean Society of London, Burlington House, Piccadilly, W.I., London, Great Britain. Linnaeus collections.

NY, The New York Botanical Garden, Bronx, NY 10458. Hagelstein collections; also J. B. Ellis, Massee, Sturgis, Brândză exsiccati, and others.

NYS, Herbarium of the New York State Museum, NY State Department of Education, Albany, NY 12224. Peck collections.

PAN, Department of Botany, Panjab University, Chandigarh, India. Thind and other Indian collections.

PH, Department of Botany, Academy of Natural Sciences, 19th and the Parkway, Philadelphia, PA 19103. Schweinitz and Bilgram collections.

TNS, National Science Museum, 3–23–1, Hyakunincho, Shinjuku, Tokyo, Japan 160. Emoto collections.

UC, Herbarium of the University of California, Department of Botany, University of California, Berkeley, CA 94720. Part of Kowalski collections.

UPS, The Herbarium, Institute of Systematic Botany, University of Uppsala, P. O. Box 541, S-751 21, Uppsala I, Sweden. E. Fries collections.

Botanical Institute of the Polish Academy of Sciences (Inst. Bot. Polsk Akad. Nauk), Warszawa, Poland [Not listed in Holmgren and Keuken, 1974]. Rostafinski, Raciborski, Krzemieniewska collections.

XII. TAXONOMY

Class **MYXOMYCETES**
Synopsis of subordinate taxa recognized, in order of treatment.

Subclass **CERATIOMYXOMYCETIDAE**

Order **CERATIOMYXALES**

Family **Ceratiomyxaceae**
Genus *Ceratiomyxa*, p. 39

Subclass **MYXOGASTROMYCETIDAE**

Order **LICEALES**

Family **Liceaceae**
Genus *Licea*, p. 42
Genus *Listerella*, p. 42

Family **Enteridiaceae** (Reticulariaceae)
Genus *Tubifera*, p. 43
Genus *Dictydiaethalium*, p. 43
Genus *Lycogala*, p. 44
Genus *Enteridium*, p. 44 (*Reticularia*)

Family **Cribrariaceae**
Genus *Lindbladia*, p. 46
Genus *Cribraria*, p. 47
Genus *Dictydium*, p. 47

Order **ECHINOSTELIALES**

Family **Echinosteliaceae**
Genus *Echinostelium*, p. 49

Family **Clastodermataceae**
Genus *Barbeyella*, p. 51
Genus *Clastoderma*, p. 51

Order **TRICHIALES**

Family **Dianemaceae**
Genus *Calomyxa*, p. 54
Genus *Dianema*, p. 54

Family **Trichiaceae**
Genus *Minakatella*, p. 56
Genus *Prototrichia*, p. 56
Genus *Perichaena*, p. 57
Genus *Oligonema*, p. 57
Genus *Calonema*, p. 57
Genus *Arcyria*, p. 58
Genus *Arcyodes*, p. 59
Genus *Cornuvia*, p. 59
Genus *Metatrichia*, p. 60
Genus *Hemitrichia*, p. 61
Genus *Trichia*, p. 61

Order **PHYSARALES**

Family **Elaeomyxaceae**
Genus *Elaeomyxa*, p. 63

Family **Physaraceae**
Genus *Protophysarum*, p. 65
Genus *Willkommlangea*, p. 65 (*Cienkowskia*)
Genus *Leocarpus*, p. 66
Genus *Physarella*, p. 66
Genus *Badhamia*, p. 67
Genus *Fuligo*, p. 68
Genus *Erionema*, p. 68
Genus *Craterium*, p. 69
Genus *Physarum*, p. 69

Family **Didymiaceae**
Genus *Diachea*, p. 72
Genus *Physarina*, p. 72
Genus *Diderma*, p. 73
Genus *Mucilago*, p. 73
Genus *Didymium*, p. 74
Genus *Lepidoderma*, p. 75

Subclass **STEMONITOMYCETIDAE**

Order **Stemonitales**

Family **Schenellaceae**
Genus *Schenella*, p. 79

Family **Stemonitaceae**
Genus *Brefeldia*, p. 80
Genus *Amaurochaete*, p. 80
Genus *Colloderma*, p. 81
Genus *Leptoderma*, p. 81
Genus *Diacheopsis*, p. 82
Genus *Enerthenema*, p. 82
Genus *Stemonitis*, p. 82
Genus *Macbrideola*, p. 83
Genus *Lamproderma*, p. 84
Genus *Comatricha*, p. 84

XIII. *Taxonomic Treatment*

Kingdom MYCETEAE: Division

GYMNOMYCOTA

Phagotrophic organisms with somatic structures devoid of cell walls. Reproduction by spores that germinate to produce amoeboid or amoeboflagellate cells.

As discussed earlier, the members of this division are regarded by many modern biologists as protists rather than fungi, but they have been studied (and hence classified) mostly by mycologists, using the botanical code of nomenclature. The system employed by Alexopoulos and Mims (**1979**) is followed here. A zoological classification (Olive, **1975**) places these organisms in the phylum Gymnomyxa of the kingdom Protista, subphylum Mycetozoa, class Eumycetozoa. There the protostelids and true slime molds are separated into subclass Protostelia and subclass Myxogastria, respectively.

The Gymnomycota may be subdivided as follows:

A. Assimilative stage an aggregation of myxamoebae (pseudoplasmodium), never a true plasmodium. Subdivision **ACRASIOGYMNOMYCOTINA**

AA. Assimilative stage a myxamoeba or a true plasmodium. Subdivision **PLASMODIOGYMNOMYCOTINA**

The subdivision Acrasiogymnomycotina comprises the "cellular slime molds." Subdivision Plasmodiogymnomycotina is divided into two classes as follows:

A. No sexual reproduction known; plasmodium, when present, minute, reticulate, with unidirectional streaming; sporophores microscopic, containing one to few spores. Class **PROTOSTELIOMYCETES**

AA. Sexual reproduction typically present; plasmodia mostly large and with reversible streaming, only in a few minute species microscopic and with irregular streaming; sporophores usually elaborate, with a few to thousands of spores. Class **MYXOMYCETES**

The class Protosteliomycetes consists of the protostelids — fairly recently discovered (Olive and Stoianovitch, **1966**), minute organisms that are invisible in the field but frequently appear in moist-chamber cultures of dry plant material or dung. Some form minute, reticulate plasmodia characterized by unidirectional streaming. In earlier years (Raper, **1973**) the protostelids were classified with the cellular slime molds in the Acrasiomycetes, but nowadays they are considered ancestral to both the cellular and the true slime molds (Olive, **1975**), and recent ultrastructural studies as well as the discovery of *Ceratiomyxella* indicate a closer relationship with the true slime molds than with the acrasid slime molds.

Class

MYXOMYCETES

Link, Handb. Gew. **3**: 405. 1833 (as suborder)*

Mycetozoa Bary, Bot. Zeitung **16**: 369. 1858 (as Mycetozoen).

Assimilative phase a multinucleate, free-living, acellular mass of protoplasm, the plasmodium, naked, amoeboid, varying from very small and giving rise to a single sporophore, to an extensive system of branching and anastomosing veins in which the active protoplasm is enclosed in an amorphous gelatinous sheath with the anterior advancing portion naked; usually immersed within the interstices of wood, bark, leaves, dung, soil, or litter and emerging to the surface before or at the time of sporulating, and capable of movement within or on the substratum, less commonly superficial from the first; under unfavorable conditions, plasmodium sometimes fragmenting into a large number of disconnected cysts or becoming transformed into a horny mass, the sclerotium, which consists of multinucleate, cell-like units capable of resuming plasmodial characteristics under favorable circumstances; reproductive phase characterized by spores or sporelike cysts, in the first subclass borne on individual stalks, in the other two subclasses borne in the interior of spore cases which usually are seated on a horny, spongy, membranous, or calcareous base, the hypothallus; spores, on germination, giving rise to one or more naked myxamoebae or flagellated swarm cells.

In the following treatment, the genera in each family are arranged in the order in which they appear in the keys.

KEY TO SUBCLASSES

A. Spores borne externally at the tips of individual hairlike stalks, on columnar, dendroid, or morchelloid sporophores **CERATIOMYXOMYCETIDAE** p. *38*
A single order, Ceratiomyxales.

AA. Spores borne in masses within various types of sporophores; peridium persistent to evanescent or only partly developed b

B. Sporophore development myxogastroid (subhypothallic); assimilative stage varied, but not an aphanoplasmodium; lime present or absent; stalk, when present, usually more or less stuffed with lime, cells, or debris **MYXOGASTROMYCETIDAE** p. *40*

BB. Sporophore development stemonitoid (epihypothallic); assimilative stage an aphanoplasmodium; lime absent (except sometimes in *Leptoderma*, a genus of doubtful affinity; see p. *81*); stalk, when present, hollow or entirely fibrous **STEMONITOMYCETIDAE** p. *76*

---

*Myxomycetes Wallroth, Fl. Crypt. Germ. **2**: 333. 1833 (as order), appeared earlier in 1833 than Link's Handbuch, but Wallroth included gasteromycete genera in it. While Link did include a few genera not now regarded as Myxomycetes, none of them are gasteromycetes, and his grouping is therefore more in accordance with present opinion than is that of Wallroth. Myxogastres Fries, Syst. Myc. **3**: 67. 1829 was published as the second suborder of his order Gasteromycetes genuini; the first suborder, Trichogastres, included mainly gasteromycetes, nearly all puffballs, and his name suggests relationships with that group. All of these names included only the endosporous forms. Since Link was the first to treat them as a distinctive group, the class name is credited to him, and Fries' name is used as the base for the subclass embracing the endogenous species only.

Subclass I.

CERATIOMYXOMYCETIDAE

Martin ex Martin & Alexopoulos. Myxomycetes 32. 1969.
CERATIOMYXOMYCETIDAE Martin in Ainsworth, Dict. Fgi. ed. 5. 497. 1961 (not validly publ.).
EXOSPOREAE Rostafinski. Vers. 2. 1873 (as Cohors).

Spores borne externally on individual spicules scattered over the surface of usually erect, often branching and anastomosing, sometimes poroid or effused, sporophores, each spore producing on germination a single, amoeboid, quadrinucleate protoplast which, after a thread phase and mitosis, is transformed into 8 haploid swarm cells.

The sporophores of *Ceratiomyxa* have in the past been regarded as branches of a massive hypothallus, partly because meiotic divisions occur in the spores, whereas for the endosporous forms, evidence until recently was that meiosis occurred in the sporangium just preceding spore formation. Recent work (see p. *15*) has shown that in those endosporous species studied, meiosis occurs in the young spores. Moreover, the spores of *Ceratiomyxa* have been demonstrated to be covered by an electron-dense layer interpreted as a counterpart of the peridium in endosporous forms (Scheetz, **1972**). Therefore, in *Ceratiomyxa*, the basal portions of the sporulating structures as well as the branches, when present, must be regarded as constituting the fructification. There is often a thin, colorless layer on the substratum under the fructification, sometimes white when dusted with spores, and the use of the term hypothallus in *Ceratiomyxa* must be restricted to that membrane. The latter may be difficult to see and sometimes appears to be completely absent.

There is a single order, family, and genus.

Ceratiomyxa fruticulosa is one of the commonest and most widely distributed Myxomycetes in the world, extremely variable and found nearly everywhere that slime molds have been collected. It is the only species which has been studied developmentally (Gilbert, **1935**; Furtado and Olive, **1971**; Scheetz, **1972**; Nelson and Scheetz, **1976**). The other two species are subtropical or tropical. It is possible that they are extreme variations of this species, but on the basis of the material available it seems proper to recognize them as distinct.

CERATIOMYXALES

Martin ex Farr & Alexopoulos, Mycotaxon **6**: 213. 1977.
Ceratiomyxales Martin, No. Amer. Fl. **1**(1): 5. 1949 (nomen seminudum).

With the characters of the subclass. A single family.

Ceratiomyxaceae

Schroeter in Engler & Prantl, Nat. Pfl. **1**(1): 15. 1889.

With the characters of the order. A single genus with three species.

Ceratiomyxa

Schroeter in Engler & Prantl, Nat. Pfl. **1**(1): 16. 1889.

Ceratium Alb. & Schw. Consp. Fung. 358. 1805. Not *Ceratium* Schrank, 1793.

Famintzinia Hazsl., Österr. Bot. Zeitschr. 27: 85. 1877.

FIGS. 1–3
Plate I

With the characters of the family.

Type species, *Isaria mucida* Pers. (=*Ceratiomyxa fruticulosa* (Müller) Macbr.)

Subclass II

MYXOGASTROMYCETIDAE

Martin, in Ainsworth, Dict. Fungi ed. 5. 497. 1961.
MYXOGASTRES Fries, Syst. Myc. 3: 3. 1829 (as suborder).
ENDOSPOREAE Rost., Versuch 2. 1873 (as Cohors).

Spores borne internally in fructifications of characteristic form, each spore producing, on germination, one or two, rarely more, swarm cells or myxamoebae; sporophore development myxogastroid (subhypothallic), the plasmodial protoplast rising internally through the developing stalk in stipitate forms; peridium continuous with stalk and hypothallus; plasmodium of various types, but not an aphanoplasmodium; calcium carbonate (lime) or calcium oxalate sometimes present.

KEY TO ORDERS

a. Spores in mass mostly pallid to brightly colored, occasionally brown to blackish, but not violaceous, by transmitted light pallid to smoky brown, but not purplish; lime absent, rarely simulated by calcium oxalate deposits on the peridium. b

b. True capillitium absent (but see *Listerella*, p. *42*); pseudocapillitium, when present, consisting of irregular, flaccid tubules or perforated membranes which may fray out into filaments. **Liceales** (p. *41*)

bb. True capillitium typically present; if absent, then sporophores minute, stalked. c

c. Fructifications consisting of minute, stalked, white or pale-colored to light brown sporangia not exceeding 1.5 mm in height, usually shorter; stalk filled with granular material; columella typically present, rarely lacking; capillitium, when present, delicate, usually not abundant. **Echinosteliales** (p. *49*)

cc. Sporophores of various forms, rarely less than 200 µm in diam, often considerably larger; columella absent; capillitium usually abundant (occasionally scanty), coarse, often sculptured. **Trichiales** (p. *53*)

aa. Spores in mass dark brown or deep violet to black (occasionally ferruginous to deep red), by transmitted light medium to dark purplish brown, rarely pallid; lime (calcium carbonate) commonly present on one or more parts of the sporophore (except in *Protophysarum* q.v.). **Physarales** (p. *62*)

LICEALES

Jahn in Engler & Prantl, Nat. Pfl. ed. 2. **2**: 319. 1928.

Capillitium lacking; pseudocapillitium present or absent; when present, in the form of irregular sculptured tubules, perforated or fraying membranes, threads, or bristles (*Tubifera bombarda*); columella absent (columellalike rods occurring in *Tubifera casparyi*); spores in mass mostly pallid to medium brown, occasionally dark brown, by transmitted light hyaline or tinted to smoky or yellow-brown, never purplish brown.

KEY TO FAMILIES

a. Fructification consisting of small, often minute sporangia or small, simple to sparsely branched, rarely effused, plasmodiocarps; neither pseudocapillitium nor dictydine granules present; spores mostly dingy to dark brown in mass and smoky by transmitted light, but sometimes bright-colored in mass, then tinted yellow or ochraceous by transmitted light. b

  b. Capillitium present, consisting of moniliform threads with angular thickenings; sporophores sporangiate, minute, black, dehiscent by preformed lobes; spores dusky. *Listerella paradoxa* (see p. *42*)

  bb. Capillitium absent. **Liceaceae** (p. *41*)

aa. Fructifications sporangiate to pseudoaethaliate or aethaliate, often large and conspicuous; spores pallid to variously colored but never smoky. c

  c. Dictydine granules lacking; fructification usually aethaliate or pseudoaethaliate; if sporangiate, then portions of peridium not persisting as a preformed net. **Enteridiaceae** (p. *43*)

  cc. Dictydine granules present; fructification mostly sporangiate, rarely united into a pseudoaethalium or an aethalium; portions of peridium in sporangiate forms persisting as a preformed net. **Cribrariaceae** (p. *46*)

Liceaceae

Rost., Versuch 4. 1873 (as Tribus)

Fructifications sporangiate, often minute, sessile or stalked, or of small and unbranched or sparsely branched (rarely netted or effused) plasmodiocarps; peridium varying from thin to thick, often becoming encrusted with a dark outer layer, sometimes clearly double; neither capillitium nor pseudocapillitium present; spores yellow-brown to reddish brown (rarely blackish) in mass, nearly colorless to smoky, yellow, reddish or smoky gray or olivaceous by transmitted light, often paler on one side, smooth to minutely warted or spinulose. Assimilative stage a protoplasmodium (except in one species of *Licea*).

With the single genus *Licea*.

The principal character of the family is the lack of capillitium or pseudocapillitium. Dehiscence may be by irregular breaking of the peridium above, where it is thinner, by a longitudinal slit, by the breaking away of the peridium in angular plates on preformed lines, or by separation of a lid. Various species have been segregated into separate genera on the basis of these characters, and method of dehiscence was used by Nannenga-Bremekamp (**1965**a) as the basis for the erection of three subgenera, *Licea*, *Orcadella* (Wing.) Nann.-Brem., and *Pleiomorpha* (also as *Pleomorphe* and

Pleismorpha) Nann.-Brem. The last of these was later raised to genus level by Dhillon (**1978**).

Rostafinski's tribe Liceaceae included both *Licea* and *Tubulina* (*Tubifera*), but more recent monographs have consistently removed *Tubifera* from the Liceaceae, leaving only *Licea* (and genera now considered synonymous).

Licea

FIGS. 4–15 *Plate* I

Schrader, Nov. Gen. Pl. 16. 1797.

Cylichnium Wallr., Fl. Crypt. Germ. **2**: 267. 1833.
Protoderma Rost., Mon. 90. 1874.
Protodermium Rost. ex Berl. in Sacc., Syll. Fung. **7**: 328. 1888.
Orcadella Wing., Proc. Acad. Phila. **41**: 280. 1889.
Protodermium Kuntze, Rev. Gen. Pl. 867. 1891.
Hymenobolus Zukal, Österr. Bot. Zeitschr. **43**: 133. 1893. Not *Hymenobolus* Dur. & Mont., 1845.
Hymenobolina Zukal, Österr. Bot. Zeitschr. **43**: 133. 1893.
Kleistobolus Lipp., Verh. Zool.-Bot. Ges. Wien **44**: Abh. 70. 1894.

With the characters of the family.

Type species, *Licea pusilla* Schrader.

As originally described, *Licea* included four species, two of which are now regarded as synonyms of *Tubifera ferruginosa*. Of the other two, *L. pusilla* is a typical example of the genus as at present delimited. *Licea variabilis* is also recognized essentially in its original application, but in its relatively large size and marked pulvinate to plasmodiocarpous habit, suggests *Perichaena*, lacking only a capillitium. It is also the only species of *Licea* known to possess a well-developed phaneroplasmodium. Occasional fruitings of *Perichaena chrysosperma* in which elaters are lacking or very sparse, approach the pulvinate forms of *L. variabilis* very closely.

The various segregates which have been given generic rank are based on single characters which are often distinctive but tend to be inconstant. Nannenga-Bremekamp's (**1965**) treatment of the genus has been most useful although, on the basis of material studied, it has not been possible to accept all of her conclusions. In their recent key to species of *Licea*, Keller and Brooks (**1977**) expanded the genus to thirty species. At least four of these are cosmopolitan (Alexopoulos, **1973**) and a few others apparently are common. Because of their minute size and drab colors, they are difficult to spot in the field, and many are known only from moist-chamber cultures.

Listerella

FIG. 61 *Plate* VI

Jahn, Ber. Deutsch. Bot. Ges. **24**: 540. 1906.

Sporangia minute, hemispherical; wall membranous, dehiscing in lobes; capillitium of slender, moniliform threads attached at base and walls; spores black in mass, dingy by transmitted light.

Type (and only) species, *Listerella paradoxa* Jahn.

This genus is essentially a Licea with what appears to be a capillitium. *Listerella paradoxa* appears indistinguishable from *Licea pusilla* and *L. minima* except for the peculiar beaded threads interspersed with the spores. Kowalski (**1968**) has suggested that the threads may represent a reduced pseudocapillitium or an extension of the peridial lobes found in some species of *Licea*. Even though the true nature of the filaments is not yet known, however, the genus seems more appropriately placed in the Liceales, as suggested by Kowalski (**1968**), rather than in the Trichiales where it has remained until now. It would seem highly unlikely that a taxon macroscopically identical with three species of *Licea* and possessing a protoplasmodium (Eliasson, **1977**b) should belong to a different order. Despite these similarities with *Licea*, however, until the nature of the filaments is elucidated, *Listerella* does not properly belong in either Liceaceae or Enteridiaceae, and its position remains doubtful.

Enteridiaceae

Farr, Mycologia **74**:339. 1982.

Reticulariaceae Rost., Versuch 6. 1873 (as Tribus).

Fructification aethaliate or sporangiate, the sporangia densely clustered, often united into a pseudoaethalium; pseudocapillitium usually present in the form of simple or branched columns which appear to represent aborted sporangia, or of bristles, or of simple to branched tubes, or of frayed or perforated membranes; spores pallid to ochraceous, olivaceous, or brown (rarely yellow) in mass, hyaline or bright yellow-brown by transmitted light, never smoky.

Since *Reticularia* was replaced by *Enteridium* (see p. 45), the long-used family name Reticulariaceae is illegitimate according to Art. 18.1, 18.3, and 64.1 of the International Code of Botanical Nomenclature.

KEY TO GENERA

a. Fructification sporangiate; sporangia clustered or united into a pseudoaethalium. b

aa. Fructification a true aethalium. c

b. Sporangiate or pseudoaethaliate with persistent sporangial walls; hypothallus massive, fibrous or spongy. **Tubifera** (p. *43*)

bb. Sporangia closely appressed into a pseudoaethalium; sporangial walls disappearing at maturity except for thickened strands at the angles, which persist as pseudocapillitial threads depending from the lids; hypothallus not massive or spongy. **Dictydiaethalium** (p. *43*)

c. Aethalium subglobose to conical or pulvinate, often on a restricted base; pseudocapillitium of colorless, branching tubes; spores pinkish, then pallid in mass. **Lycogala** (p. *44*)

cc. Aethalium pulvinate on a broad base; pseudocapillitium of frayed or perforated membranes; spores brown, yellow, or olivaceous in mass. **Enteridium** (p. *44*)

Tubifera

J. F. Gmelin, Syst. Nat. **2**: 1472. 1791.

FIGS. 16–20
Plate II

Tubulifera Jacq., Misc. Austr. **1**: 144. 1778.

Tubulina Pers., Neues Mag. Bot. **1**: 91. 1794.

Alwisia Berk. & Br., J. Linn. Soc. **14**: 86. 1873.

Siphoptychium Rost., Mon. App. 32. 1876.

Sporangia cylindric or ellipsoid, free or connate on a usually thick, spongy hypothallus and then often forming a pseudoaethalium, the walls membranous, persistent; dehiscence apical; capillitium lacking; pseudocapillitium present or absent, when present, as bristles arising from walls of sporangial cavity or as a more or less branching columellalike protrusion which may represent an abortive sporangium; spores bright yellow-brown, spiny or reticulate.

Type species, *Stemonitis ferruginosa* Batsch (=*Tubifera ferruginosa* (Batsch) J. F. Gmelin, Syst. Nat. **2**: 1472. 1791.)

The genus contains five species, of which one is tropical, ana one known only from India. The type species and, possibly, *T. microsperma* are cosmopolitan.

Dictydiaethalium

Rost., Versuch 5. 1873.

FIG. 21
Plate II

Clathroptychium Rost., Mon. 225. 1875.

Ophiuridium Hazsl., Österr. Bot. Zeitschr. **27**: 84. 1877.

Fructification a pseudoaethalium, composed of numerous cylindrical sporangia closely compacted into a palisade layer and angular by pressure, the thinner portions of the walls disappearing at maturity, leaving the thicker regions formed by the angles depending from the thickened caps of the sporangia as threads forming a pseudocapillitium, the caps united to form a continuous bullate or tessellate cortex; spores ochraceous, brown, or olivaceous, less commonly red or yellow in mass, paler by transmitted light.

Type species, *Reticularia plumbea* Schum.) Fries (=*Dictydiaethalium plumbeum* (Schum.) Rost. in A. Lister, Mycet. 157. 1894).

The fructification is commonly referred to as an aethalium. Baker (**1933**) showed that until maturity it is composed of closely compacted cylindrical sporangia, each with a complete wall which is thickened at the junctions. At maturity, the thinner portions of the walls disappear, leaving the thickened angular corners depending as threads from the caps of the individual sporangia and immersed in the general mass of the spores as a pseudocapillitium, the threads not reaching the base. To call this a pseudoaethalium rather than an aethalium is thus largely a matter of definition.

Besides the cosmopolitan type species, one other species has been described from New Caledonia.

Lycogala

FIGS. 22–25
Plate II

ADANS., FAM. PL. **2**: 7. 1763.

Galeperdon Wiggers, Prim. Fl. Holsat. 108. 1780.
Diphtherium Ehrenb., Sylvae Myc. Berol. 26. 1818.
Dermodium Rost., Mon. 284. 1875. Not *Dermodium* Link, 1815.
Antonigeppia Kuntze, Rev. Gen. Pl. **3**: 443. 1898.
Verrucosia Teng, Contr. Biol. Lab. Sci. Soc. China **8**: 124. 1937.

Fructification a globose, conical, or pulvinate aethalium, resembling a puff ball; cortex varying from a thick, firm, crustose shell or a rather spongy layer to a delicate membrane, nearly smooth or bearing wartlike protuberances which may be divided into tessellated chambers; pseudocapillitium of branched or simple tubes, variously sculptured to nearly smooth, sometimes penetrating the cortex. Spores often pinkish at first, changing to gray or ochraceous in mass, nearly colorless by transmitted light.

Type species, *Lycoperdon epidendrum* L. (=*Lycogala epidendrum* (L.) Fries, Syst. Myc. **3**: 80. 1829).

The genus *Lycogala* was established by Micheli (**1729**). His illustration of *L. flavofuscum* (as *L. griseum, majus*), *pl.* 95, below, is particularly good. On the same plate he also illustrated three phases of *L. epidendrum* and an additional species which appears to represent *Leocarpus fragilis*.

The spores of all species are small and those of most are delicately reticulated; the differences in the pseudocapillitium are not striking but seem fairly constant. The species are distinguished mainly by size, shape, and the nature of the cortex and its markings (Martin, **1967**). The latter are formed from the slime sheath encasing the plasmodium (Eliasson and Sunhede, **1980**).

The genus consists of four classic species, of which *L. epidendrum* and *L. exiguum* are cosmopolitan; *L. flavofuscum* and *L. conicum* are widely distributed. Two additional species have been named from India.

Enteridium

FIGS. 26–31
Plate III

Ehrenberg, Jahrb. Gewächsk. **1**: 55. 1819.

Reticularia Bull., Champ. Fr. 83. 1791; not *Reticularia* Baumg., Fl. Lips. 569–570. 1790.

Licaethalium Rost., Vers. 4. 1873.

Liceopsis Torr., Bull. Soc. Portug. Sci. Nat. **2**: 63. 1908.

Fructification an aethalium; pseudocapillitium arising from the base and varying from perforated membranes or membranes fraying out into threads to entirely threadlike, the threads often united into a reticulation; spores brown in mass, usually more or less reticulate when free, warted on exposed surfaces when clustered.

Type species, *Enteridium olivaceum* Ehrenb.

The distinction between *Reticularia* and *Enteridium* was based on the character of the pseudocapillitium, which in the former was described as fraying out into threads, in the latter as consisting of perforated membranes. In its extreme manifestation this is striking, but there are too many intermediate expressions between the extremes to justify using this character as the basis for separate genera.

According to Farr (**1976**a), Bulliard's 1789 treatment was not definitive for the genus *Reticularia*, whereas his 1791 usage constituted a later homonym of the lichen genus *Reticularia* Baumg., necessitating replacement of the generic name in the Myxomycetes by *Enteridium*. Nannenga-Bremekamp (**1979**, App.), however, judging that Bulliard's 1789 illustrations could be nothing other than *R. lycoperdon*, retained Bulliard's generic name.

Cribrariaceae

Rost., Versuch 5. 1873 (as Tribus)

Sporophores sporangiate, usually stalked (in *Lindbladia* mostly sessile and then often merged into pseudoaethalia or aethalia); capillitium lacking; all parts of sporophore (including spores) bearing minute, typically dark granules (dictydine granules); peridium in *Lindbladia* continuous or nearly so or, if netted, interstices rarely fugacious; in the other genera netted over the entire surface or over the upper part, the interstices fugacious, so that at maturity the peridium is represented either by a complete net or net above and a calyculus below; spores yellow, brown, red, or purple in mass, pale or bright-colored by transmitted light.

The dictydine granules, usually conspicuous except in *Lindbladia*, the lack of a pseudocapillitium or capillitium, and the persistent surface net in *Cribraria* and *Dictydium* make most members of this family easy to recognize as such.

KEY TO GENERA

a. Sporangia usually closely aggregated on an extensive, often thick and spongy hypothallus, the walls often adherent or fused, forming a pseudoaethalium or an aethalium, rarely scattered; net lacking or scantily developed, and peridium rarely if ever dehiscent between the meshes; dictydine granules few and concolorous with membranes. **Lindbladia** (p. *46*)

aa. Sporangia usually free, aggregated or scattered; net always present, usually well-developed; hypothallus delicate; dictydine granules numerous, usually darker than spores. b

  b. Threads of net short, meeting at thickened or expanded nodes. **Cribraria** (p. *47*)

  bb. Main threads of net stout, longitudinal, subparallel at least below, connected by very delicate transverse threads; peridium sometimes netted above. **Dictydium** (p. *47*)

Lindbladia

FIG. 32
Plate III

Fries, Summa Veg. Scand. 449. 1849.

Fructifications typically aethaliate but varying from sessile or rarely substipitate, clustered to densely massed sporangia through pseudoaethaliate; peridium relatively thick, without a surface net or with a delicate surface net on some or all of the constituent sporangia, but not fugacious between meshes; dictydine granules present but concolorous with membranes and usually inconspicuous; hypothallus extensive, firm, often thick and more or less spongy; spores dark olivaceous brown in mass.

Type (and only) species, *Lindbladia tubulina* Fr.

This genus is distinctly intermediate between the Enteridiaceae and the Cribrariaceae, but grouped with the latter on the basis of the dictydine granules and the occasional presence of a net on the surface of the upper part of the peridium. The more sporangiate fruitings approach those of *Cribraria argillacea* rather closely; in the specimens which are regarded as pseudoaethalia, the sporangia, although united, retain their identity. In some developments, however, there is no evidence that sporangia were differentiated before the spores were formed; such sporophores are regarded as true aethalia.

Cribraria

Persoon, Neues Mag. Bot. 1: 91. 1794.

FIGS. 33–53
Plates III–V

Hypothallus thin, delicate, or indiscernible; fructifications sporangiate, globose or pyriform, usually stalked; peridium thickened above and often below in netlike fashion, fugacious at maturity between the meshes of the net, leaving only the netted portion and frequently a shallow to deep calyculus, from which the net arises; veins of net short, meeting at nodes which are usually expanded and sometimes notably thickened; dictydine granules present on calyculus and net and usually on spores.

Type species, *Cribraria rufescens* Pers. (=*C. rufa* (Roth) Rost., Mon. 232. 1875.)

Cribraria is for the most part a well-marked genus. Sporangiate phases of *Lindbladia tubulina* do resemble certain phases of *C. argillacea*, but in the former the net, if present, is superimposed on the persistent peridium, whereas in the latter the peridium is much thinner and falls away from the interstices of the net. Some fruitings of *Dictydium mirabile* are quite *Cribraria*-like, as are occasional fruitings of *Dictydium cancellatum*, particularly those which have developed under alpine conditions. For these reasons, Nannenga-Bremekamp (**1962**) proposed to unite the two genera. Martin (**1962**) studied the same problem independently and decided they were better kept separate. There is no serious disagreement about the facts involved; it is merely a question as to how they are best interpreted. For the present it seems advisable to maintain the separation, as explained on p. *48*.

The species often present great difficulty. All of the characters used to distinguish them tend to be inconstant and to vary with the maturity of the collection and the conditions under which it may have ripened. This is well illustrated by the presence or absence of a calyculus. Probably all species have a peridium which tends to persist at the base longer than above. In the majority of species this is fixed. In others, of which *C. microcarpa* and *C. intricata* are the commonest examples, the basal portion, in well-developed specimens, tends to disappear at maturity, but in specimens which have been checked by drying before full maturity a calyculus may be present, although often rudimentary or incomplete. Some collections may show a range in this character, with sporangia in exposed areas having a calyculus while those in more sheltered portions lack it. The distinctions between flat and thickened nodes is useful and fairly constant, but intermediate phases do occur. This character is best seen in mounts in which the nodes may be observed in profile at the margins of the sporangia. Color of sporangium and of spore mass is distinctive in some species, but not in all. Size of spores is useful for a few species and the same is true for color and size of plasmodic (dictydine) granules.

This large genus of ca. two dozen species is worldwide in distribution. The most common and widely reported species are *Cribraria intricata*, *C. languescens*, *C. microcarpa*, *C. tenella*, and *C. violacea* (the latter especially from moist-chamber cultures).

Dictydium

Schrader, Nov. Gen. Pl. 11. 1797.

Heterodictyon Rost., Versuch 5. 1873.

FIGS. 54–56
Plate VI

Sporangiate, stalked, the sporangia globose or subglobose, often umbilicate above or below, or both; peridium delicate, usually evanescent above and more tardily so below except for the net and a basal portion which may remain as a calyculus; dictydine granules prominent, usually dark, densely aggregated on all parts of net, calyculus (when present), and spores; net composed of stout longitudinal ribs connected by delicate transverse filaments, especially below, the upper portion sometimes more or less netted (as in *Cribraria*), without thickened nodes.

Type species, *Dictydium umbilicatum* Schrader (=*D. cancellatum* (Batsch) Macbr., No. Amer. Slime-Moulds 172. 1899.)

Although he did not specifically say so, Schrader apparently made the absence of a calyculus the distinctive feature by which he distinguished *Dictydium* from *Cribraria*, and it was not until Rostafinski's treatment of 1875 that the current application of these names was adopted. Nannenga-Bremekamp (**1962**) merged *Dictydium* with *Cribraria*, pointing out that neither presence nor absence of a calyculus is constant, and that the net of *Dictydium* merges into that of *Cribraria*. There is much merit in this disposition, but similar problems occur in other genera. In the present case, *Dictydium* is at least as constant as many of these other genera, its only common species is readily recognized by the naked eye, and it seems more convenient to retain such a familiar and well-known genus.

Besides the common and ubiquitous type species, two others are known, one from mountainous regions (*D. mirabile*) and one only from Australia (*D. rutilum*).

ECHINOSTELIALES

Martin, Mycologia **52**: 127. 1961.

Sporangia globose, stalked, minute, bright-colored to light or medium brown; columella globose, cylindrical, fusiform, or conical, sometimes lacking; stalk stuffed with granules; capillitium varying from a complete open net or a system of branching and anastomosing threads to a few branched threads, or lacking; spores white, pale pink or yellow to ochraceous or brown in mass, nearly colorless by transmitted light, sometimes with thickened areolae in the wall; assimilative stage a protoplasmodium.

In the monograph of Martin and Alexopoulos (**1969**) the order included only the genus *Echinostelium*. *Clastoderma* and *Barbeyella*, in keeping with tradition, were maintained in the Stemonitales. In their comments on *Clastoderma*, however, the authors noted the differences in sporangium development and type of plasmodium, and recommended the transfer of that genus to the Echinosteliaceae, as suggested by Alexopoulos (**1969**). Two years later (Alexopoulos and Brooks, **1971**) the family Clastodermataceae was formally established and circumscribed to comprise *Clastoderma* and *Barbeyella*. Further historical details on the classification of these genera may be found in the last-cited paper.

The present classification scheme of the Echinosteliales assembles what are thought to be closely related genera in the light of our present state of knowledge.

KEY TO FAMILIES (slightly modified from Alexopoulos and Brooks, **1971**)

a. Spore mass white to cream-colored, gray, yellow, or pink, rarely pale pinkish brown; peridium delicate, evanescent at an early stage (except in one species of *Echinostelium*), though often persisting basally as a collar. **Echinosteliaceae** (p. *49*)

aa. Spore mass brown; peridium persistent either as a whole or in fragments which cling to the tips of the capillitium. **Clastodermataceae** (p. *51*)

Echinosteliaceae

Rost., Versuch 7. 1873 (as Tribus).

Peridium delicate, evanescent except for a persistent basal collar in some species (persistent in *Echinostelium arboreum*); sporangia rarely exceeding 50 µm in diam, up to 500 µm high; columella variously shaped or lacking; spores white or lightly tinted in mass, sometimes with a thin area or with one or more areolae (thickened areas) in the wall.

With the single genus, *Echinostelium*.

Echinostelium

Bary, A. de, in Rost., Versuch 7. 1873. FIG. 60

?*Heimerleia* Höhnel, Ann. Mycol. **1**: 391. 1903. *Plate* VI

With the characters of the family.

Type species, *Echinostelium minutum* Bary.

The genus *Echinostelium* is now one of the better known genera of the Myxomycetes (Alexopoulos, **1958**, **1960**a, **1961**; Ing, **1965**; Haskins, **1971**; Keller and Brooks, **1976**b; Whitney, **1980**). Its traditional position in the Stemonitales has always been anomalous, and as early as 1961, Martin gave reasons for placing it in a separate order. His arguments were fortified by the characteristic spore characters, which are common to several species and unlike those of any other Myxomycetes. These spores are smooth in outline, but under high powers of the microscope the wall is seen to be characterized by more or less circular, platelike areas, representing contact points between spores (Whitney, **1980**). The protoplasmodia are minute and amoebalike, each giving rise to a single sporangium.

There are several records of the collection of *E. minutum* in the field; however, most of our knowledge of that species, and all that we know of the others, is based on material which has appeared in moist chambers, mostly on bark.

As a result of the resurgence in interest and increased activity in moist-chamber culture of Myxomycetes, additional species of *Echinostelium* have been described by various collectors in rapid succession, culminating in a painstaking monograph by Whitney (**1980**) who recognized thirteen species. Some of these are separated on rather minute differences, but considering the size of the sporulating structures, that is understandable. The reliability of some of the diagnostic characters utilized to separate species remains to be determined in the future.

Heimerleia is listed in modern monographs as synonym of *Echinostelium*. The original description (Höhnel, **1903**) of this "totally hyaline" organism cites a thin peridium, full-length tapering columella, sporangial diam. of 70–120 μm, height of 120 μm, and no capillitium. These characters are borne out in Höhnel's later illustration (*fig. c*, **1914**). It is unlikely that the taxon is *E. minutum*, nor does it fit any of the other known species of *Echinostelium*. It may have referred to a mucor.

Another possible synonym, *Endodromia* Berk., is noted in the literature (Lister, **1925**; Martin and Alexopoulos, **1969**). This genus was described and illustrated as minute, stipitate, hyaline sporangia with persistent peridium dehiscing into "little granular portions," percurrent columella, a branched capillitium, and spores containing a mobile nucleus. The genus is typified only by Berkeley's figures, since no type specimen was preserved. Although the description does not necessarily exclude *Echinostelium* (considering the optics of the times), more probably *Endodromia vitrea* refers to a mucor, quite possibly similar to, or identical with, Höhnel's *Heimerleia*.

Clastodermataceae

Alexop. and Brooks, Mycologia **63**: 926. 1971.

Peridium wholly or partly persistent, either remaining below as a cup and splitting above into petaloid lobes, or fragmenting into small, scalelike pieces which cling to the tips of the capillitial threads; sporangia usually 100–200 μm in diam; stalk dark; spores brown in mass.

KEY TO GENERA (modified from Alexopoulos and Brooks, **1971**).

a. Peridium tough, wholly persistent, splitting above into petaloid lobes which remain attached to the cuplike basal portion. **Barbeyella** (p. *51*)

aa. Peridium delicate, evanescent except for small, scalelike fragments which cling to the tips of the capillitial threads. **Clastoderma** (p. *51*)

Barbeyella

Meylan, Bull. Soc. Bot. Genève II. **6**: 89. 1914.

FIG. 185
Plate XX

Sporangia stipitate, columellate, globose, dehiscent by petaloid lobes or platelets along preformed lines (Kowalski and Hinchee, **1972**); stalk dark, filled with dark granules; columella an extension of the stalk, giving rise at apex to 7–10 simple or sparsely branched capillitial threads, one or two to a lobe; spores dark.

Type (and only) species, *Barbeyella minutissima* Meylan.

Barbeyella minutissima resembles *Licea operculata* as noted by Meylan, but differs not only in lack of a lid, but in the presence of a columella and a scanty but distinct capillitium. The color, as mentioned by G. Lister (**1925**), is that of a *Lamproderma*; the columella, with capillitium originating at the tip, and the firm peridium also show similarity with that genus. Jarocki (**1931**) reported finding *Barbeyella minutissima* in numerous localities in eastern Poland, in damp situations on the under-surface of decaying spruce logs. The description of the plasmodial stage is based on his account; he also describes its development. Emoto (**1933**) reported the species from Japan, Lakhanpal and Mukerji (**1981**) from India, and Curtis (**1968**) from Oregon. In this country, this minute slime mold is also known from Washington (state), where it is reported to be abundant on leafy liverworts in the mountains (Kowalski and Hinchee, **1972**). The last-mentioned authors also recognized the close relationship between *Barbeyella* and *Clastoderma*, and the relative remoteness of these genera from *Lamproderma*, based on stalk structure, plasmodium type, and capillitium characters of these three genera.

Clastoderma

Blytt, Bot. Zeit. **38**: 343. 1880.

FIG. 184
Plate XX

Orthotricha Wing., J. Mycol. **2**: 125. 1886. Not *Orthotrichium* Hedw., 1789.

Wingina Kuntze, Rev. Gen. Pl. **1**: 875. 1891.

Sporangia globose, stipitate; peridium breaking up at maturity into rounded or angular fragments which remain attached to the tips of the capillitium; columella short or lacking; capillitium arising from the apex of the columella or the base of the sporangium, consisting of branching and anastomosing threads bearing at the free tips the peridial platelets; spores brown.

Type species, *Clastoderma debaryanum* Blytt.

Clastoderma debaryanum is common and cosmopolitan. It occurs in moist chambers as well as in the field and is easily recognized by its small size and usually a small droplet near the middle of the stalk, which divides the latter into two distinct regions. *Clastoderma pachypus* differs from *C. debaryanum* in the much thicker stalk and the somewhat longer columella. Lakhanpal and Mukerji (**1981**) recognize two additional species.

TRICHIALES

Macbride, N. A. Slime-Moulds, ed. 2. 237. 1922.

Plasmodiocarpous or sporangiate, rarely pseudoaethaliate, sessile or stalked; columella never present; spores in mass typically bright-colored, white to yellow, orange, or red, by transmitted light hyaline to bright-colored (blackish brown in mass and dusky by transmitted light in *Listerella*, here included in the Liceales); capillitium threadlike, solid or tubular, smooth or sculptured, free or attached.

KEY TO FAMILIES

a. Capillitium of solid or nearly solid threads, attached to base and often to sporangial walls, never united into a net. **Dianemaceae** (p. 53)

aa. Capillitium of tubular threads, free or attached to base of sporangium, often united into a net. **Trichiaceae** (p. 55)

Detailed light-microscope and ultrastructural observations on capillitial structure in the Trichiales (Kowalski, **1968**; Keller et al, **1973**) have cast doubt on the taxonomic validity of the Dianemaceae. Kowalski (**1968**) moved *Prototrichia* to the Trichiaceae because he found the capillitium of that genus to be hollow. Keller et al (**1973**) obtained similar results for the capillitium of *Minakatella*, causing transfer of that genus likewise to the Trichiaceae. These studies have caused some investigators to predict a merger of the two families (Collins, **1979**). In view of the fact, however, that Ellis et al (**1973**) found entirely solid capillitium in *Calomyxa* and lack of a true lumen in *Dianema* (as opposed to the truly hollow trichiaceous types of capillitium) a union of the two families at this time would be premature.

Dianemaceae

Macbride, N. Am. Slime-Moulds 180. 1899 (as Dianemeae).

Plasmodiocarpous or sporangiate, rarely short-stipitate; peridium usually single or with a granular outer layer; capillitium composed of threads which are either solid or with a restricted lumen, smooth or with minute sculpturing (moniliform in *Listerella*, here included in the Liceales), either coiled and hairlike or nearly straight, attached to the base of the fructifications and usually to the peridium, simple or sparsely branched but never forming a net.

The Dianemaceae of Macbride, **1922**, is in modern usage essentially the same as the Margaritaceae of the Lister monographs. However, since *Margarita* A. Lister, when published, was a later homonym of *Margarita* Gaudin it cannot, under Art. 18 and 64 of the ICBN (Stafleu et al, **1978**), be used to give its name to a family of Myxomycetes.

The enigmatic genus *Listerella* traditionally has been included with hesitation in this family because of its peculiar capillitium-like filaments, and is therefore included in the following key. For reasons explained on p. *42*, however, the genus is transferred to the Liceales.

KEY TO GENERA

a. Sporangia black, minute, dehiscent by preformed lobes; capillitium with angular thickenings, appearing moniliform; spores dusky. **Listerella** (p. *42*) aa. Sporangia not black; dehiscence irregular; capillitium not moniliform; spores bright-colored. b

b. Capillitial threads slender, hairlike, coiled, with few attachments to the peridium. **Calomyxa** (p. 54)

bb. Capillitial threads relatively stout, nearly straight, with many of the tips attached to the peridium. **Dianema** (p. 54)

Calomyxa

FIG. 62
Plate VI

Nieuwl., Am. Midl. Nat. **4**: 335. 1916.

Margarita A. Lister, Mycet. 203. 1894. Not *Margarita* Gaudin, 1829.

Sporophores sporangiate, sessile or rarely stalked, globose to pulvinate, varying to plasmodiocarpous; peridium membranous, translucent or covered with granules; capillitium of simple or sparsely branched, coiled or flexuous, slender, solid threads, minutely sculptured, attached at base and often to the peridium.

Type species, *Physarum metallicum* Berk. (=*Calomyxa metallica* (Berk.) Nieuwland, Am. Midl. Nat. **4**: 335. 1916.

Calomyxa metallica is a widely distributed species that appears fairly commonly in moist-chamber cultures of bark from dead or living trees and is then often in the form of isolated sporangia or short plasmodiocarps. Field collections tend to be more densely aggregated and often more or less coalesced sporangia, sometimes approaching a pseudoaethalium. Plasmodiocarps may appear in the same development with sporangia. The variability of the species has resulted in the description of the varieties *intermedia* and *plasmodiocarpa*, neither of which appears to have merit. The variety *microspora* is a synonym of *Dianema depressum* (Kowalski, **1975**a).

In the second and third editions of the Lister monograph (**1911, 1925**), *Licea incarnata* and its derivatives in other genera are cited as possible synonyms of this species. This is an unlikely disposition, as explained under *Arcyodes incarnata* (p. 59).

Calomyxa synsporos, the second species, is known only by the type collection from the Venezuelan Andes.

Dianema

FIGS. 63–65
Plate VI

Rex, Proc. Acad. Phila. **43**: 397. 1891.

Lamprodermopsis Meylan, Bull. Soc. Vaud. Sci. Nat. **46**: 56. 1910.

Plasmodiocarpous to sporangiate, sessile or substipitate; peridium membranous to cartilaginous; capillitium of smooth or obscurely sculptured, simple or forked, slender threads, attached to the base and mostly to the peridium; spores pallid or yellow, sometimes pinkish at first, free or united into clusters.

Type species, *Dianema harveyi* Rex.

There is frequent reference in the literature to the pink color of the spores, both in mass and as seen under the microscope. This is not apparent in dried specimens, but it is not unlikely that, as in the familiar *Lycogala epidendrum* and other species, the pink color may be present in freshly matured fructifications.

The genus *Dianema* is a small one of approximately a half-dozen species which seem to be confined to the cool (and often mountainous) regions of the world.

Trichiaceae

Rost., Versuch 14. 1873 (as Tribus).

Sporangiate, sessile or stalked, or plasmodiocarpous; capillitium tubular, sculptured in characteristic fashion or nearly smooth, consisting of simple, branched threads called "elaters," or united into a net, free or attached at the base; spores white or bright-colored in mass, hyaline, colorless to bright yellow or red by transmitted light.

The genera here recognized are distinguished mainly by the characters of the capillitium, chiefly by the sculpturing on the threads and whether these are coiled, united into a net, or broken up into free elaters. Some of these distinctions are not always sharp and may show much variation even within a species. It is quite possible that some of the smaller genera will prove to be superfluous and that the limits of others may have to be altered, but the classification here adopted has proved to be, on the whole, workable, and is maintained with some revision.

KEY TO GENERA

a. Capillitium coiled, simple or sparsely branched, sometimes dividing penicillately above or anastomosed at bases and tips. b

aa. Capillitium reticulate or of simple to branched elaters, not coiled. c

b. Capillitial threads smooth to spiny on one side, flattened, with bulbous ends, intercalary swellings, and expanded junctions often filled with bacteria. **Minakatella** (p. *56)*

bb. Capillitial threads often spirally twisted about each other, becoming subdivided near the top into penicillate tips that are attached to the upper wall, sometimes anastomosing into a network at bases and tips, usually at least partly marked with spiral bands. **Prototrichia** (p. *56)*

c. Capillitium bearing spines, cogs, or rings, sometimes nearly smooth or more or less reticulate, or with well-marked to faint, poorly defined spirals intermixed with other markings. d

cc. Capillitium bearing 2–6 well-defined smooth to spiny spiral bands. i

d. Capillitium of free elaters, these usually short, simple or sparsely branched; if long, rarely forming a complete net. e

dd. Capillitium of long, profusely branched and anastomosing threads, typically united into a net. f

e. Elaters warted, spiny, or nearly smooth, or minutely annulate; sporophores sporangiate to plasmodiocarpous or, if densely clustered, not heaped; peridium rather thick, usually impregnated with granular material, appearing double, rarely with excreted calcium oxalate. **Perichaena** (p. *57)*

ee. Elaters bearing faint and irregular spirals or nearly smooth; sporangia densely aggregated, usually heaped; peridium thin, membranous, often iridescent. **Oligonema** (p. *57)*

f. Capillitium marked as in *Oligonema*, but threads united into an incomplete net. **Calonema** (p. *57)*

ff. Capillitium variously marked, but rarely with spirals and then with calyculus and fugacious upper peridium. g

g. Peridium usually fugacious above the persistent calyculus; capillitium often strongly elastic. **Arcyria** (p. *58)*

gg. Peridium tending to be persistent, especially below, but not forming a morphologically distinct calyculus; capillitium somewhat elastic. h

h. Capillitium bearing warts or spines; sporangia small, sessile, heaped. **Arcyodes** (p. *59)*

hh. Capillitium bearing prominent coarse rings. **Cornuvia** (p. *59)*

i. Peridium cartilaginous, thick, shining, opening by a preformed lid; elaters notably spiny. **Metatrichia** (p. *60)*

ii. Peridium membranous or thickened by accretion and then dull, opening irregularly or in lobate fashion or, if by a lid, then both cup and lid membranous; elaters spiny or smooth. j

j. Threads of capillitium united into an intricate net, with few free ends. **Hemitrichia** (p. *61*)

jj. Threads of capillitium broken into relatively short, unbranched or sparsely branched elaters, hence free ends numerous. **Trichia** (p. *61*)

Minakatella

G. Lister, J. Bot. **59**: 92. 1921.

Sporangia sessile, densely clustered or united into a pseudoaethalium; peridium membranous, often iridescent; capillitium consisting of coiled tubular threads with unevenly thickened walls and with bulbous ends, intercalary swellings, and expanded, flattened, axial junctions with two or three divergent threads, which often are plugged with bacteria or protoplasmic debris, roughened or spinose to smooth; spores brightly colored. (This description is partly based on that of Keller et al, **1973**.)

Type (and only) species, *Minakatella longifila* G. Lister.

G. Lister (**1921**) described the capillitium as tubular and regarded the genus as related to *Perichaena*, differing from the latter only "in the smooth capillitium and the aethalioid habit." This was refuted by later workers, particularly Martin and Alexopoulos (**1969**), who considered *Minakatella* closely related to (if not synonymous with) *Calomyxa*. Recent light- and ultramicroscopic studies by Keller et al (**1973**) confirmed Lister's description. These investigators found similarities as well as differences between *Minakatella* and *Calomyxa* on the one hand, and between *Minakatella* and *Perichaena* and *Arcyria* on the other. Finding the capillitium to be hollow rather than solid as in *Calomyxa*, they transferred *Minakatella* to the Trichiaceae. Their electron micrographs clearly show the tubular nature of the threads and the presence of bacterial remains, as in the capillitium of *Arcyria* (Mims, **1969**). Their recommendation to maintain *Minakatella* as a separate genus is being followed here, especially in view of the findings by Ellis et al (**1973**) regarding the capillitium of *Calomyxa* (see p. 53) and pending further studies on capillitial structure in *Perichaena*. Keller et al's (**1973**) allusion to the similarities of the capillitium of *Calomyxa* to that of *Dianema*, and personal preliminary (light-microscope) examinations suggest that *Calomyxa* is more closely related to *Dianema* than to *Minakatella* and there seems to be little if any reason to keep *Calomyxa* separate from *Dianema*. The key characters used in the monographs appear to be of degree rather than kind.

Prototrichia

FIG. 66
Plate VI

Rost., Mon. App. 38. 1876.

Sporangiate, sessile, rarely short-stipitate or subplasmodiocarpous; peridium thin, transparent; capillitium of nearly solid threads, smooth or sculptured with faint to distinct spiral bands, often twisted about each other in spirals, attached at the base of the sporangium and becoming subdivided above, the penicillate tips attached to the upper wall; spores at first pinkish then brown in mass, pinkish then yellow by transmitted light.

Type (and only) species, *Trichia metallica* Berk. (=*Prototrichia metallica* (Berk.) Massee, J. Roy. Microsc. Soc. 1889: 350. 1889.)

In some specimens the capillitium shows little or no evidence of spirals; in most, the spirals occur on at least a good proportion of the threads and often on practically all of them. Scanning electron micrographs by Ellis et al (**1973**) and Rammeloo (**1978**) show the ornamentations to be scarcely protruding spiral bulges of the capillitial thread, hence their often faint appearance. The presence of spiral markings and the discovery by Kowalski (**1968**), confirmed by Ellis et al (**1973**), that the capillitial threads are not entirely solid, support placement of *Prototrichia* in the Trichiaceae rather than in the Dianemaceae, where it formerly was classified.

In North America, *Prototrichia metallica* is known mostly from the western mountain regions. It is also widely distributed in Europe.

Perichaena

Fries, Symb. Gast. 11. 1817.

FIGS. 67–73
Plate VII

Pyxidium S. F. Gray, Nat. Arr. Brit. Pl. **1**: 580. 1821.

Stegasma Corda, Ic. Fung. **5**: 58. 1842.

Ophiotheca Currey, Quart. J. Micr. Sci. **2**: 241. 1854.

Sporangiate to plasmodiocarpous; peridium usually double, the outer layer granular, rarely encrusted with calcium oxalate, sometimes poorly developed, the inner layer membranous, closely attached to the outer; capillitium of simple or branched tubular threads, slightly roughened to warty or spiny or, in one species, minutely annulate, but not bearing spirals; spores yellow, minutely warted or spinulose.

Type species, *Perichaena populina* Fr. (=*P. corticalis* (Batsch) Rost., Mon. 293. 1875.)

Perichaena differs from *Trichia* and *Oligonema* in lacking spiral bands on the capillitium. *Oligonema*, furthermore, has a single membranous, shining peridium, whereas that of all except two *Perichaena* species is double. (In some species, however, the two layers are united, or one may be poorly developed, so that the double nature of the peridium is not always evident.)

The calcareous peridial deposits occasionally found in some species of *Perichaena* are calcium oxalate (Lister, **1925**; Jahn, **1928**; Schoknecht and Keller, **1977**), not calcium carbonate as in the Physarales.

Perichaena contains about nine species, of which three (*P. chrysosperma*, *P. corticalis*, and *P. depressa*) are common and ubiquitous. *Perichaena vermicularis* is also cosmopolitan but less common than the other three.

Oligonema

Rost., Mon. 291. 1875.

FIGS. 74–76
Plate VII

Fructifications sporangiate, the sporangia usually densely crowded, tending to be superimposed or heaped; peridium thin, membranous; capillitium of short or long, simple or branched elaters, nearly smooth or obscurely sculptured with spirals and sometimes with spines, warts, or rings; spores yellow.

Type species, *Trichia nitens* Libert, 1834 (=*Oligonema schweinitzii* (Berk.) Martin, Mycologia **39**: 460. 1947.) Not *T. nitens* Pers., 1796.

While usually the sporangia are massed, they may be more or less scattered, particularly at the margins of extensive colonies, in the two common species, *O. flavidum* and *O. schweinitzii*. In both of these species the elaters may vary from very short to rather long. Too little is known of *O. fulvum* to generalize in this respect.

Although two of the three species of *Oligonema* are rather widely distributed, the genus is not one of the most commonly encountered ones.

Calonema

Morgan, J. Cinc. Soc. Nat. Hist. **16**: 27. 1893.

FIG. 77
Plate VIII

Sporangia sessile, subglobose or irregular to pulvinate, crowded, sometimes superimposed; peridium thin, shining, dehiscing irregularly; capillitium of branching threads arising from the base and united into a network, the surface reticulately sculptured and marked with irregular rings and spirals; spores yellow.

Type species, *Calonema aureum* Morgan.

The genus differs from *Oligonema* only by having a capillitial network instead of elaters. The capillitial markings appear to be intermediate between those of *Oligonema* and *Perichaena*.

Martin and Alexopoulos (**1969**) suggested merging *Calonema* and *Oligonema* with *Perichaena* on the basis of capillitial similarity. This similarity is far more pronounced in *Calonema luteolum* than in *C. aureum*. On the basis of capillitial markings alone, certain species of *Perichaena, Oligonema, Calonema luteolum*, and *Arcyria annulifera* certainly seem sufficiently similar as to be considered congeneric. If one adds other characters, however, the iridescent, golden-yellow sporophores of *Calonema* are indistinguishable from those of *Oligonema*, and also resemble those of *Cornuvia* and *Trichia lutescens*. Most species of *Perichaena* have dark, dull, double or thickened walls, *Oligonema* has elaters instead of a network, and *Arcyria annulifera* has typically arcyrioid sporangia.

The capillitial markings of *C. aureum* are highly variable even within one sporangium, as pointed out previously (Martin and Alexopoulos, **1969**). Even on one strand they may be in the form of spiral bands on one part, and gradually merge through increasingly more densely spaced spirals to rings on another portion of the same strand. In this character the type species approaches *Cornuvia*.

Calonema aureum is not common, but apparently widely distributed throughout the U.S.A.

The second species, *C. luteolum*, is cited by its author (Kowalski, **1969**) from numerous collections — all on cow dung in the Sacramento Valley, California. A portion of the type, as well as one specimen collected two years later in Mexico on a decaying twig, are deposited in BPI. Although the sporophores certainly look like those of *Calonema* or *Oligonema*, the capillitium lacks spiral markings. It appears beaded or possibly ringed and very irregular in outline, with many expanded junctions, strongly resembling that of some species of *Perichaena* and that of *Arcyria annulifera*. Braun and Keller (**1976**) suggested that *Calonema luteolum, Perichaena quadrata*, and *P. liceoides* may be conspecific.

Arcyria

FIGS. 78–86 *Plate* VIII

FIGS. 87–96 *Plate* IX

Wiggers, Prim. Fl. Holsat. 109. 1780.

Nassula Fries, Summa Veg. Scand. 456. 1849.

Arcyrella (Rost.) Racib., Rozp. Akad. Umiej. **12**: 80. 1884.

Heterotrichia Massee, Mon. 139. 1892.

Sporangia subcylindric, ovoid or subglobose, stipitate or sometimes sessile by a joint; peridium thin, fugacious above, typically separating by a definite line of dehiscence just above the base, the lower portion remaining as a persistent funnel-, cup-, or saucer-shaped calyculus, in some species the margin between the persistent base and the fugacious upper part not distinct; stalks often packed with round bodies larger than the spores; capillitium netted, elastic, frequently expanding to more than twice the original height of the sporangium after dehiscence, either attached to both base and sides of the calyculus and tending to remain in place or merely to the center of the calyculus at the junction of the stipe and then breaking away freely, variously ornamented with half-rings, cogs, warts, spines, rings, reticulations, or inconspicuous (occasionally well-developed) spiral bands; spores concolorous in mass, hyaline to bright-colored by transmitted light.

Type species, *Clathrus denudatus* L. (=*Arcyria denudata* (L.) Wettst., Verh. Zool.-Bot. Ges. Wien **35**: Abh. 535. 1886.)

Most of the species of *Arcyria* are readily recognized as belonging to the genus. Two species, *A. leiocarpa* and *A. stipata*, because of the spiral bands on the capillitium, have often been referred to *Hemitrichia*. In all but this character, however, they are closer to *Arcyria* and have been included in this genus in most recent monographs.

Rostafinski (**1875**, pp. *268, 274*) divided the genus into two subgenera: *Clathroides*, characterized by firm attachment of the capillitium to the calyculus, and *Arcyrella*, with the capillitium loosely attached. Raciborski (**1884**) used *Arcyrella* as a genus, without formally raising its rank. The difference, while a fairly useful key character, is neither distinct enough nor sufficiently constant to justify its use as a basis for generic segregation.

Robbrecht (**1974**) utilized the size of the round bodies in the stalk as a diagnostic character in his synoptic key to the Belgian species of *Arcyria*.

Arscyria is a variant spelling.

This distinctive and rather large genus encompasses, among its 22 or so species, some of the most common and best-known Myxomycetes in the world. *Arcyria cinerea, A. denudata, A. incarnata, A. insignis, A. nutans*, and possibly *A. pomiformis* are cosmopolitan and easy to recognize in the field with a handlens.

Arcyodes

O. F. Cook, Science **15**: 651. 1902.

FIG. 97
Plate IX

Lachnobolus Fries, Summa Veg. Scand. 457. 1849. Not *Lachnobolus* Fries, 1825.

Sporangia distinct, clustered, sessile or short-stalked, often heaped; peridium single, membranous, persistent, at least below, as a deep, irregularly lobed calyculus; capillitium a loose, irregular, inelastic network, the threads spiny, warty, or somewhat reticulate, arising from the base and attached at numerous points to the peridium; spores pallid.

Type (and only) species, *Licea incarnata* Alb. & Schw. (=*Arcyodes incarnata* (Alb. & Schw.) O. F. Cook, Science **15**: 651. 1902.)

As explained by Martin and Alexopoulos (**1969**), *Lachnobolus* Fries, 1849, is a later homonym of *Lachnobolus* Fries, 1825, and is invalid. Since the generic name *Lachnobolus* was used in different senses it had to be rejected under Art. 69 of the ICBN. Cook decided that the only way out of the confusion was to rename the genus. His suggestion was ignored in the various subsequent monographs until Martin (**1949**) accepted the genus *Arcyodes*.

The type of *Arcyodes* must be *Licea incarnata* Alb. & Schw. It must be admitted that the figure accompanying the original diagnosis is not at all convincing. The description, on the other hand, allows for enough variation from the illustrated concept to warrant the assumption that it refers to the species under discussion. Furthermore, in Fries' (**1829**) treatment of *Perichaena congesta* and *P. incarnata* the description of the latter sounds more likely as applying to the *Arcyodes* at hand than does the depiction of *P. congesta*. Lister's (**1925**, p. 252) suggestion that *Licea incarnata* may have referred to *Margarita metallica* (=*Calomyxa metallica*) seems highly improbable.

Arcyodes incarnata is known from Europe and North America but appears to be rather rare. Often, North American specimens originally referred to this species were incorrectly determined, many belonging to *Arcyria occidentalis*. The latter species, because of its relatively inelastic capillitium, also was formerly included in *Lachnobolus*. The capillitium is quite different.

Cornuvia

Rost., Versuch 15. 1873.

FIG. 98
Plate X

Fructifications sporangiate or plasmodiocarpous, sessile; capillitium not elastic, consisting of a network of flaccid threads, with free ends, marked with coarse rings.

Type (and only) species, *Arcyria serpula* Wig. (=*Cornuvia serpula* (Wig.) Rost. in Fuckel, Jahrb. Nass. Ver. Nat. **27–28**: 76. 1873.)

Cornuvia serpula shows affinities with both *Arcyria* and *Hemitrichia*. Martin and Alexopoulos (**1969**) suggested the possibility of returning it to *Arcyria* or, alternatively, to expand *Cornuvia* to include *Arcyria annulifera*.

The only collection in BPI, a scant but well-developed specimen from India, consists of bright yellow, sometimes iridescent plasmodiocarps with delicate membranous peridium, and capillitium and spores fitting the description well. The general aspect and habit, color, and particularly the coarsely reticulate spores, suggest *Hemitrichia*. The capillitium does not, but neither does it resemble that of any *Arcyria* species. Recent observations on the isotype (and only) collection of *A. annulifera* in the BPI herbarium bear out the annotations by Martin and Alexopoulos (**1969**, pp. *123–124*): The scanty, moldy specimen shows well-defined cups, slender, scanty, nonelastic capillitium with faint rings, and spores marked with a few scattered warts. In other words, *A. annulifera* has the sporangia and spores of a typical *Arcyria* and the delicate, beaded (ringed) capillitium of certain *Perichaena* species. This evidence does not favor a merger of *Cornuvia serpula* with *Arcyria annulifera*.

Cornuvia serpula, apparently a rare slime mold, has been reported from Europe, India, and Michigan.

The type of *Cornuvia anomala* is a battered but clearly recognizable specimen of *Arcyodes incarnata*, as annotated by E. Baskerville. Contrary to the statements in Martin and Alexopoulos (**1969**), both specimen and slides are preserved in the Iowa collection, which now is part of National Fungus Collections (BPI).

Metatrichia

FIG. 121
Plate XII

Ing, Trans. Brit. Mycol. Soc. **47**: 51. 1964.

Sporophores sporangiate, sessile or stalked, often united into clusters or large aggregates forming pseudoaethalia; peridium iridescent or lustrous, double, consisting of a membranous inner layer closely appressed to a more or less cartilaginous outer layer, dehiscent by a preformed operculum, or inoperculate and then differentiated into a thickened calyculus and a membranous upper portion with wart-like thickenings; elaters free, unbranched, usually long and twisted, bearing prominent, strongly developed spines; capillitium and spores deep orange-red to crimson in mass.

Type species, *Metatrichia horrida* Ing.

In both the type species and the extremely common *M. vesparium*, the inner layer of the peridium is smooth and the outer layer suggests an incrustation, but is also shining. The two parts are so closely united that the double nature of the wall is not apparent except in mounts examined under the microscope. In the paratype of *M. horrida*, both layers of the peridium are delicate and membranous, so that the peridium as a whole is thinner and more fragile than that of *M. vesparium*. In all species there is a tendency for the elaters to be bent back in the middle, with the two halves coiled about each other; occasionally that character, the reddish color, and the spiny capillitium may be found in species of *Trichia*. Lakhanpal and Mukerji (**1977**) expanded *Metatrichia* to four species. The type species is known only from Nigeria, and *M. paragoga* only from the West Indies, but *M. vesparium* is one of the most universally distributed and common slime molds the world over. It is recognized in the field by the distinctive clusters of sporangia which, after opening of the lids, resemble miniature red or blackish paper-wasp nests. In *M. paragoga*, dehiscence is not operculate; the peridium forms a thickened calyculus, and the upper portion is membranous with embedded, thickened warts. The species was transferred from *Hemitrichia* on the basis of its *Metatrichia*like capillitium. The fourth species placed in *Metatrichia* by Lakhanpal and Mukerji (**1977**) is *M. arundinariae*. This species does not have a *Metatrichia*-type capillitium and its original placement in *Trichia* (Rammeloo, **1973**), near *T. floriformis*, appears to be correct.

Hemitrichia

Rost., Versuch 14. 1873.

Hyporhamma Corda, Ic. Fung. **6**: 13. 1854. (*nomen confusum*).

Hemiarcyria Rost., Mon. 261. 1875.

FIG. 113 *Plate* XI

FIG. 367 *Plate* XLI

FIGS. 114–120 *Plate* XII

Fructifications sporangiate, either stalked or sessile, or plasmodiocarpous; peridium membranous or subcartilaginous, usually persistent below as an irregular calyculus, usually thinner and more or less fugacious above; stalk, when present, solid or filled with sporelike vesicles or amorphous material; capillitium of tubular threads united more or less completely into an elastic net, with or without free ends, and ornamented with two or more usually conspicuous spiral bands; spores red, orange, or yellow in mass, bright or pale by transmitted light.

Type species, *Trichia clavata* Pers. (=*Hemitrichia clavata* (Pers.) Rost. in Fuckel, Jahrb. Nass. Ver. Nat. **27–28**: 75. 1873.)

In certain species the capillitium may occasionally be broken up into free elaters, as mentioned in the Lister monographs, and care should be taken to make allowance for that fact before describing such forms as new. These species are in some degree intermediate with *Trichia*.

The genus *Hemitrichia* contains about a dozen species, of which *H. serpula* is cosmopolitan. *Hemitrichia calyculata* (formerly *H. stipitata*) is common in the tropics and subtropics, and widely distributed in many temperate regions, although it is not known from the Netherlands (Nannenga-Bremekamp, **1974**), nor listed from Belgium by Rammeloo (**1978**). *Hemitrichia clavata* ranges throughout the Temperate Zone, but does not occur in truly tropical environments.

Trichia

Haller, Hist. Stirp. Helv. **3**: 114. 1768.

FIGS. 99–105 *Plate* X

FIGS. 106–112 *Plate* XI

Sporophores sporangiate, stipitate or sessile, or subplasmodiocarpous; peridium membranous or cartilaginous; capillitium elastic, of free, simple or sparsely branched elaters, marked with 2–5, rarely more, spiral bands; spores yellow, yellow-brown, or reddish in mass, hyaline or tinted by transmitted light.

Type species, *Trichia ovata* Pers. (=*T. varia* (Pers.) Persoon, Neues Mag. Bot. **1**: 90. 1794.)

As currently delimited, this genus contains 14 species, the most common of which are *Trichia favoginea* (including *T. affinis* and *T. persimilis* which some European authors recognize as distinct species), *T. floriformis*, and *T. varia*.

PHYSARALES

Macbride, N. Am. Slime-Moulds ed. 2. 22. 1922.

Spores black, deep purplish, or violaceous brown in mass, deep purplish brown to violaceous by transmitted light; lime (calcium carbonate) usually visible, often abundant, in any, some, or all parts of the sporophore except the spores (except in *Protophysarum* and family Elaeomyxaceae); capillitium of threadlike or tubular filaments throughout, or bearing limy nodes; assimilative stage a phaneroplasmodium.

The Physarales typically are distinguished by the usually abundant lime in some and often all parts of the fructification. The recently discovered genus *Protophysarum* does not display any visible lime, but was included in the Physarales on the strength of its mitochondrial calcium granules (Blackwell, **1974**). It is not known whether or not calcium is present at the ultrastructural level in *Elaeomyxa*. The latter is placed in this order tentatively, for reasons stated on p. *63*.

Diachea has a limy stalk, columella, and hypothallus and for that reason was included in this order in the Lister and Hagelstein monographs. Others (Morgan, **1894**; Macbride and Martin, **1934**; Martin, **1949**; Martin and Alexopoulos, **1969**) placed the genus in the Stemonitales because of its resemblance to *Lamproderma*. More compelling modern reasons for classifying *Diachea* in the Physarales include evidence, based on stalk structure, indicating a subhypothallic method of sporophore development (Blackwell, **1974**) and reports of phaneroplasmodia for *D. leucopodia* (Rammeloo, **1978**).

Leptoderma has been classified in the Physarales by some, and in the Stemonitales by others. Its uncertain taxonomic position is discussed on p. *81*.

KEY TO FAMILIES

a. Lime absent; wax or oil present in one or more parts of the sporophore. **Elaeomyxaceae** (p. *62*)

aa. Wax absent; lime usually present in one or more parts of the sporophore. b

  b. Capillitium calcareous, usually intricate. **Physaraceae** (p. *64*)

  bb. Capillitium noncalcareous (rarely bearing aggregations or trabeculae, i.e. rods, of crystalline lime. **Didymiaceae** (p. *71*)

Elaeomyxaceae

Hagelst. ex Farr and Keller, Mycologia **74**: 857. 1982.
Elaeomyxaceae Hagelst., Mycologia **34**: 594. 1942. *Nomen nudum*.

Sporophores stalked or sessile, limeless; peridium membranous; wax or oil nodules included within the stalk (when present) and sometimes in the columella, capillitium, and peridium; capillitium composed of dark purplish, branching and anastomosing threads.

A single genus with two species.

Hagelstein included *Elaeomyxa* in the order Amaurochaetales, suborder Amaurochaetinae (which comprised the Stemonitales and Echinosteliales of present-day systems) but, recognizing its lack of affinity with other genera in that suborder, proposed a separate family for it. His family designation, however, was not accepted in subsequent literature, even by himself. In his monograph (**1944**) he listed the genus in family

"Stemonitidaceae," where it has remained, although as an acknowledged misfit. With the advent of newer classification systems emphasizing sporophore development, recent authors such as Alexopoulos (**1973**), Keller and Candoussau (**1973**), and Blackwell (**1974**) concluded that the stuffed stalk structure in *Elaeomyxa* precluded close affinity with the Stemonitales, but rather is indicative of a subhypothallic mode of development. On the basis of these findings, Keller (**1980**) transferred the genus to the Physarales and keyed it in the Didymiaceae because of its superficial resemblance to *Diachea*. Farr and Keller (**1982**) retained *Elaeomyxa* in the Physarales and, deciding it does not fit into either the Physaraceae or the Didymiaceae, validated Hagelstein's family to accommodate it.

Elaeomyxa

Hagelst., Mycologia **34**: 593. 1942.

FIGS. 130–131
Plate XIII

With the characters of the family.

Type species, *Diachea miyazakiensis* Emoto (=*Elaeomyxa miyazakiensis* (Emoto) Hagelst., Mycologia **34**: 593. 1942)

The wax in the stalk and wall is clearly seen only when the specimens are wet or mounted. That in the capillitium of *E. miyazakiensis* is in the form of often nodular swellings on the threads.

The genus consists of the type species and *E. cerifera*. The type species is represented in America by ample and excellent material from Ontario, Canada. It looks like a limeless, robust *Diachea leucopodia*. In both species of *Elaeomyxa*, the stout black stalks are partly to completely filled with dark debris and, sometimes, waxy droplets. The hypothallus ranges from individual patches to venous strips connecting two or three sporangia; it has hyaline edges but becomes darker toward the inside, gradually merging into the base of the stalk. *Elaeomyxa cerifera*, not known from this continent, sometimes has a waxy collar at the apex of the stalk.

Physaraceae

Rost. Versuch 9. 1873 (as Tribus).

Capillitium netted, limy, very rarely nearly limeless, composed of calcareous tubes of nearly uniform diameter, of calcareous nodes connected by slender, hyaline tubules, or of a combination of these and other characters; peridium usually limy; spores black, deep violaceous, or dark gray in mass, deep purplish brown to violaceous brown or pale violaceous by transmitted light.

The generic distinctions in the Physaraceae are not always sharp. Five of the genera here recognized are monotypic, but even though the characters by which they are distinguished are not necessarily fundamental, they are readily recognizable to the naked eye and remarkably constant. *Physarum*, by far the largest genus in the Myxomycetes, displays a wide range of forms, and many attempts have been made to divide it into smaller genera, but none of the propositions have proved to be practicable. It merges into *Badhamia*, as noted under that genus, and also into *Craterium* and *Fuligo*, but in practice these genera are usually readily recognizable and it seems best to maintain them until we possess further knowledge of their developmental morphology.

There are many plasmodiocarpous species, but the majority are sporangiate. However, many sporangiate species may, under certain circumstances, sporulate as plasmodiocarps. In such cases there are nearly always sporangia present and there is often a complete series of transitions from the sporangia to the plasmodiocarps.

Fuligo is the only consistently aethaliate genus. Even so, occasional fructifications of *Fuligo* approach the sporangial condition while a few species of *Physarum*, notably *P. gyrosum*, may in their sporophores suggest aethalia, but the resemblance is at most superficial.

KEY TO GENERA

a. No visible lime present; sporangia minute (up to 150 μm in diam), stipitate, suggesting *Lamproderma* or *Macbrideola*. **Protophysarum** (p. 65)

aa. Lime typically present; sporophores considerably larger. b

  b. Capillitium duplex, i.e. composed of two distinct systems. c

  bb. Capillitium essentially homogeneous. e

c. Primarily plasmodiocarpous, but sometimes forming pulvinate sporangia or massed into a pseudoaethalium; capillitium of flat nodes massed transversely into limy plates, connected with a nearly limeless network of slender tubes bearing numerous, often hooked spines. **Willkommlangea** (*Cienkowskia*) (p. 65)

cc. Primarily sporangiate or, if plasmodiocarpous, then usually accompanied by sporangia; capillitium not spinose. d

  d. Sporangia obovoid to ellipsoid or broadly cylindrical; peridium smooth, shining; capillitium a limy network, connected with and interpenetrating a limeless net of flattened tubules. **Leocarpus** (p. 66)

  dd. Sporangia deeply introverted, thimblelike, rarely plasmodiocarpous; peridium rough; capillitium composed of stout calcareous spines (trabeculae) arising from the inner wall, and a network of slender threads bearing a few calcareous nodes. **Physarella** (p. 66)

e. Capillitium a network of calcareous tubes of nearly uniform diameter; limeless connecting tubules few or none. **Badhamia*** (p. 67)

ee. Capillitium a network of hyaline limeless tubules with calcareous nodes at many or all of the junctions. f

---

*Keller and Brooks (**1976**a) removed one species with greatly flattened sporophores and simple capillitium, *Badhamia ainoae*, to form the type of a new genus, *Badhamiopsis*. Reduced capillitium of this type, consisting of calcareous tubes, may also be found in some very flat fructifications of *Badhamia affinis*.

f. Fructification an aethalium; pseudocapillitium present, often more conspicuous than the capillitium. **Fuligo** (p. 68)

ff. Fructification sporangiate or plasmodiocarpous, rarely approaching aethaliate; pseudocapillitium lacking. g

g. Plasmodiocarpous, cylindrical, pendent, often anastomosing to form a 3-dimensional net. **Erionema** (p. 68)

gg. Sporangiate or plasmodiocarpous, rarely pendent; plasmodiocarps, when anastomosing, forming a 2-dimensional net. h

h. Sporangiate; dehiscence more or less circumscissile or by a preformed lid, the lower portion of the peridium persisting as a deep, usually well-defined calyculus. **Craterium** (p. 69)

hh. Sporangiate to pseudoaethaliate or plasmodiocarpous, rarely somewhat aethalioid; dehiscence irregular or lobate, rarely areolate or by longitudinal fissure; lower portion of peridium remaining as at most a shallow, irregular cup. **Physarum** (p. 69)

Protophysarum

Blackw. & Alexop., Mycologia **67**: 33. 1975.

Sporangia minute, stipitate, limeless as observed macroscopically or by light microscopy; peridium membranous, delicate, iridescent, persistent; stalk stuffed with granular material; columella absent; capillitium arising from the occasionally pulvinate sporangial base, forming a loose network of flattened hyaline or tinted tubules often expanded at the junctions; spores dark in mass, violaceous brown by transmitted light; plasmodium a minute phaneroplasmodium.

Type (and only) species, *Protophysarum phloiogenum* Blackw. & Alexop.

As indicated by the authors of this genus (Blackwell, **1974**; Blackwell and Alexopoulos, **1975**), *Protophysarum phloiogenum* resembles a minute *Lamproderma* but is excluded from the Stemonitales by its subhypothallic development, phaneroplasmodium, and "split" type spore germination. Because of its minute sporophores and plasmodium the genus is thought to be a possible evolutionary link between the Physarales and the Echinosteliales. It was placed in the former order by virtue of the presence of mitochondrial calcium granules (Blackwell, **1974**), because the capillitium forms a *Physarum*like network, and the plasmodium, although minute, is a true phaneroplasmodium.

Willkommlangea

Kuntze, Rev. Gen. Pl. **3**: 875. 1891.

FIG. 200
Plate XXII

Cienkowskia Rost., Versuch 9. 1873. Not *Cienkowskia* Regel & Rach, 1858, nor *Cienkowskya* Solms, 1867.

Fructification primarily plasmodiocarpous, the plasmodiocarps often massed to form a pseudoaethalium, or less commonly broken up into pulvinate sporangia, irregularly dehiscent; peridium cartilaginous, more or less densely encrusted with lime; capillitium duplex, of angular, flattened nodes massed transversely into plates tending to divide the plasmodiocarp into segments, and slender, anastomosing threads forming a loose or dense net, bearing a few calcareous nodes and numerous short, sharp-pointed, often uncinate branchlets; spores dark in mass.

Type (and only) species, *Physarum reticulatum* Alb. & Schw. (=*Willkommlangea reticulata* (Alb. & Schw.) Kuntze, Rev. Gen. Pl. **3**: 875. 1891.)

Martin (**1966**) pointed out and Martin and Alexopoulos (**1969**) acknowledged that *Cienkowskia* Rost. is a later homonym for *Cienkowskia* Regel & Rach and *Cienkowskya* Solms (both obscure names in flowering plants) but retained Rostafinski's name because of its time-honored and universal application in the Myxomycetes. Since at least one, and possibly both, of the earlier names were validly published, this practice violates Art. 64 (sections 1 and 2) of the ICBN (Stafleu, **1978**). Therefore, either *Cienkowskia* Rost. must be conserved or Kuntze's name substituted.* Since only one species is involved (and a valid combination already exists for it), conservation hardly seems warranted and the latter course is here adopted. This change should not be too drastic to be easily assimilated.

Willkommlangea reticulata, a universally distributed and moderately common species, is extremely variable but rather readily recognizable in all its forms. The red spots characteristic of many fructifications, especially the paler ones, are described in the Lister monograph (**1925**) as "waxy" but that probably refers only to their appearance, not their nature. After the rupture of the peridium, the contents tend to emerge as elongate strands of capillitium and spores and then the platelike appearance of the limy capillitium is very striking. The inner wall is thicker at the base and often persists after the spores and capillitium have been dissipated. When the wall is bright-colored, the bases become very conspicuous.

Leocarpus

FIG. 201
Plate XXII

Link, Ges. Nat. Freunde Berlin Mag. 3: 25. 1809.

Tripotrichia Corda, Ic. Fung. **1**: 22. 1837.

Sporophores sporangiate; peridium brittle, of three layers: a thin cartilaginous outer layer, a thick, calcareous middle layer, and a membranous inner layer to which the capillitium is attached; capillitium duplex, consisting of a network of limy tubes resembling that of *Badhamia*, connected with but mostly distinct from a nearly limeless network of flattened tubules expanded at the junctions; columella lacking; pseudocolumella often present; spores black in mass.

Type (and only) species, *Diderma vernicosum* Pers. (=*Leocarpus fragilis* (Dicks.) Rost., Mon. 132. 1874.)

Common and cosmopolitan, *Leocarpus fragilis* is a variable but very distinctive species. The smooth, shining, extremely fragile sporangia, often clustered on twigs, suggest a mass of insect eggs. Under a hand lens or binocular microscope a broken sporangium with its prominent limy capillitium showing sharply against the black spore mass resembles a *Badhamia*. Under such circumstances the hyaline portion of the capillitium is difficult to see. Also, the extreme fragility of the peridium extends to the capillitium so that unusual care must be used in making microscopic mounts. Despite these difficulties and the great variation in color and shape, the species is readily recognizable at sight.

Lister (**1925**) speaks of the peridium as composed of two layers, regarding the limy shell as deposited inside the outer layer and part of it. Rostafinski (**1875**, *pl. 6, fig. 93*) and Baker (**1933**, *pl. 3, fig. 20*) regard it as triple. This seems correct and is easily demonstrated.

Physarella

FIG. 202
Plate XXII

Peck, Bull. Torrey Bot. Club **9**: 61. 1882.

Sporophores sporangiate, cylindrical or cupulate, introverted, hence thimble- or bell-shaped, borne on a hollow stalk, sometimes sessile or plasmodiocarpous; peridium firm, membranous, more or less encrusted with

---

*I thank D. Nicholson for providing information on the validity of the names in question.

lime; capillitium duplex, consisting of stout trabeculae borne on the inner surface of the exterior walls and penetrating into the interior, and a delicate network of nearly limeless tubules bearing a few fusiform calcareous nodules; spores dark in mass.

Type (and only) species, *Physarella mirabilis* Peck (=*P. oblonga* (Berk. & Curtis) Morgan, J. Cinc. Soc. Nat. Hist. **19**: 7. 1896.)

The nomenclatural history of this slime mold was reviewed by Martin and Alexopoulos (**1969**) and brought up to date by Farr (**1976**b).

This striking species, when fully developed, looks like nothing else. The thimble-shaped sporangium and, after its opening, the strongly limy trabeculae protruding from the inner surface of the outer peridium, and the tubular inner peridium jutting out as a pseudocolumella make it unmistakable. Plasmodiocarpous fruitings are not rare, but are usually accompanied by the typical sporangiate forms. Even when this is not the case, the characteristic spines are conspicuous.

A white variant with yellow stalks and white plasmodium was collected from two localities in Jamaica, W. I. (Farr, **1957**) and from Pakistan (Thind and Rehill, **1958**). One of the Jamaican strains was grown in culture by Alexopoulos (**1964**b) and proved to be consistent under such conditions, suggesting that it is an albino mutant. This is a case where a subspecific name is justified, and this organism may properly be cited as *P. oblonga* f. *alba*. G. Lister (**1925**, p. 73) cited *P. lusitanica* Torr. as a doubtful synonym. *Physarella oblonga* has now been cultured extensively and variations similar to those stressed by Torrend (**1908**, p. 113 [p. 173 of 1909 reprint]) have been observed in such cultures, thus removing the doubt.

Badhamia

Berk., Trans. Linn. Sco. **21**: 153. 1853.

FIGS. 203–206 *Plate* XXII

FIGS. 207–215 *Plate* XXIII

FIGS. 216–218 *Plate* XXIV

FIG. 366 *Plate* XLI

Sporophores sporangiate, sessile or stalked, varying to somewhat plasmodiocarpous; peridium thin, varying from nearly limeless to thickly encrusted; capillitium a network of calcareous tubules, sometimes with a few, and sometimes with many nodes and hyaline, limeless tubules, then approaching *Physarum*; stipe, when present, varying from membranous and little more than an extension of the hypothallus, to stout and well formed; columella absent; spores black in mass, free or adherent in clusters.

Type species, *Sphaerocarpus capsulifer* Bull. (=*Badhamia capsulifera* (Bull.) Berk., Trans. Linn. Soc. **21**: 153. 1853.)

As here delimited, *Badhamia* merges imperceptibly into *Physarum*. As originally defined by Berkeley it was restricted to species with clustered spores. Berkeley believed that each cluster was originally encased in a vesicle, which disappeared at maturity. Of the six species recognized by Berkeley, three species, *B. capsulifera*, *B. utricularis*, and *B. nitens*, plus *B. versicolor*, *B. papaveracea*, and *B. populina* constitute a coherent generic group. Of the remaining species, some have a typical badhamioid capillitium in some collections while in others which are regarded as representing the same species, the capillitium is very uneven, with large angular nodes connected by slender limy strands; sometimes some of the latter are limeless, as in *Physarum*. Furthermore, the spores are free. It has been suggested that some of these species should be transferred to *Physarum*, but only after detailed study of ample material (Martin and Alexopoulos, **1969**). Carter and Nannenga-Bremekamp (**1972**) emphasized the capillitium as the distinguishing character in separating *Badhamia* and *Physarum*, but neither clustering of spores nor tubular capillitium should be considered alone. As stated earlier (Farr, **1976**b), "typical species of *Badhamia* combine clustered spores with a uniformly tubular capillitium and seem to constitute a reasonably well-defined nucleus."

As currently circumscribed, *Badhamia* is one of the larger myxomycete genera, including about twenty species.

Fuligo

FIGS. 219–223
Plate XXIV

Haller, Hist. Stirp. Helv. **3**: 110. 1768.

Aethalium Link, Ges. Nat. Freunde Berlin Mag. **3**: 24. 1809.

Aethaliopsis Zopf, Pilzth. 149. 1885.

Fructification aethaliate, occasionally subplasmodiocarpous, consisting of interwoven and poorly defined tubes, each with a calcareous wall, rarely forming masses suggesting densely compacted sporangia; outer portion sterile, forming a fragile cortex, this sometimes nearly lacking; basal layer a membranous hypothallus, the intermediate portion containing spores, capillitium, and limy walls (pseudocapillitium) derived from the plasmodial tubes; capillitium of hyaline, tubular threads connecting the lime knots, often rather scanty; spores dark in mass.

Type species, *Mucor septicus* L. (=*Fuligo septica* (L.) Wiggers, Prim. Fl. Holsat. 112. 1780.)

Fuligo is very close to *Physarum* and some fruitings, particularly of *F. septica*, *F. intermedia*, and *F. cinerea* approach very closely and sometimes completely to densely massed sporangia.

The type species, *F. septica*, is one of the commonest and most widely distributed of Myxomycetes. Its extraordinary variability in size, shape, and color is reflected in the numerous names which it has received. The cortex may be very thick or sparse or even lacking, in which case the sporophores appear to be densely clustered and anastomosing sporangia on a common hypothallus; all, however, have the small, minutely warted, rather pale spores and there seems to be no way to separate these forms into coherent subgroups. The varieties *candida*, *violacea*, *flava*, and *rufa* of R. E. Fries, Svensk Bot. Tidskr. **6**: 744. 1912, were all validly published, but do no more than name the color involved. The varieties *cinnamomea* and *laevis* were irregularly and probably invalidly published. Some, at least, of the specimens referred to the var. *violacea* are in an early stage of attack by the hypocreaceous fungus *Nectria violacea* (=*Nectriopsis violacea*). Unless cultural studies can demonstrate that some of these variations are due to more than response to conditions under which the plasmodium developed and sporulated, they should be disregarded. Skupienski (**1926**) argued at length that *Fuligo rufa* should be recognized as a distinct species, citing differences in plasmodia and ecological relationships as well as in morphology. It is quite conceivable that further study, preferably involving comparative cultures, may confirm his opinion.

Morgan (**1896**) pointed out that the brief diagnosis of *Mucor septicus* L., 1763, probably referred only to the plasmodial stage. However, Wiggers, in adopting Linnaeus' epithet, noted the rapid change to a dark mass of spores.

Fuligo cinerea with its thin, white aethalia and usually darker, ellipsoid spores is another common and cosmopolitan species, often found on piles of rotting straw, manure, and plant detritus. *Fuligo intermedia* appears to be common in the western states. The other two species are not as well known.

Erionema

FIG. 224
Plate XXIV

Penzig, Myxom. Buitenz. 36. 1898. Not *Erionema* Maire, 1906 (error for *Eriomena*).

Plasmodiocarps sometimes sessile but usually pendent on slender stalks, simple or branched, often anastomosing freely and forming a complex 3-dimensional network; capillitium elastic, of numerous colorless tubules, most of the junctions unexpanded and limeless, but bearing a relatively small number of calcareous nodes.

Type (and only) species, *E. aureum* Penzig.

The genus is close to *Physarum* and *Fuligo*. The plasmodiocarps of *Physarum bogoriense*, for example, have been observed suspended from the substrate by slender filaments much as in *Erionema*. Ecorticate aethalia of *Fuligo septica* may approach *Erionema* (Lister and Lister, **1904**; Martin and Alexopoulos, **1969**), as did some tropical specimens reported by Farr (**1969**) in Dominica, W. I. None of these are known to form a three-dimensional network, however. Thus, the single species is distinctive and the genus may well be maintained pending further study.

Erionema aureum is not known from the Western Hemisphere.

Craterium

Trent. in Roth, Catalecta Bot. **1**: 224. 1797.

FIGS. 225–229
Plate XXV

Cupularia Link, Handb. **3**: 421. 1833.

Scyphium Rost., Mon. 148. 1874.

Iocraterium Jahn, Hedwigia **43**: 302. 1904.

Sporangia cyathiform or cylindric to globose, stalked, or occasionally sessile (rarely approaching plasmodiocarpous); peridium cartilaginous, more or less encrusted with lime, the lower portion tending to persist as a deep calyculus; dehiscence circumscissile or irregular at the apex or by a preformed lid; columella rarely present; capillitium of hyaline, threadlike tubes connecting calcareous nodes (rarely of badhamioid, limy tubules), the latter often aggregated in the center to form a pseudocolumella; spores dark in mass, varying from deep rose or purple to black.

Type species, *Craterium pedunculatum* Trent. (=*C. minutum* (Leers) Fr., Syst. Myc. **3**: 151. 1829.)

Craterium is distinguished by the long-persistent calyculus. The species with distinct lids are readily recognizable; those in which the lid is not clearly delimited nearly always have a cap of different texture from the rest of the peridium and it is that which is shed at dehiscence.

The genus is close to *Physarum*, and some species of the latter have a persistent, cuplike base, but *Craterium* in general is well characterized and is worthy of maintaining as a convenience, if no more.

Seven species are currently recognized, one of which, *C. obovatum*, was classified in *Badhamia* until recently. Most species are widely distributed, *C. leucocephalum* being probably the most common one. *Craterium paraguayense*, with striking purple sporangia, is mostly tropical to subtropical and known in the U.S. only from Florida, Louisiana, and Texas.

Physarum

Pers., Neues Mag. Bot. **1**: 88. 1794.

Angioridium Grev., Scot. Crypt. Fl. *pl. 210*. 1827.

Trichamphora Jungh., Crypt. Java 12. 1838.

Claustria Fries, Summa Veg. Scand. 451. 1849.

Tilmadoche Fries, Summa Veg. Scand. 454. 1849.

Crateriachea Rost., Versuch 11. 1873.

Cytidium Morgan, J. Cinc. Soc. Nat. Hist. **19**: 8. 1896.

Sporophores sporangiate to plasmodiocarpous, rarely almost aethaliate or pseudoaethaliate; peridium single or double, calcareous (rarely nearly limeless); stalk, when present, usually tubular and translucent, or stuffed with lime or dark amorphous material, sometimes with lime on the exterior surface only; columella or pseudocolumella often present; capillitium a network of hyaline tubules connecting calcareous nodes, attached to the base and to the peridium; lime in the peridium, capillitium, and stalk in the form of amorphous granules, rarely subcrystalline; spores black or dark brown in mass, violet-brown or violaceous by transmitted light.

FIGS. 230–233 *Plate* XXV
FIGS. 234–242 *Plate* XXVI
FIGS. 243–250 *Plate* XXVII
FIGS. 251–259 *Plate* XXVIII
FIGS. 260–269 *Plate* XXIX
FIGS. 270–278 *Plate* XXX
FIGS. 279–286 *Plate* XXXI
FIGS. 287–294 *Plate* XXXII
FIGS. 295–301 *Plate* XXXIII

Type species, *Physarum aureum* Pers. (=*P. viride* (Bull.) Pers., Ann. Bot. Usteri **15**: 6. 1795.)

In occasional collections of various species, the lime may be deposited on the peridium in the form of crystalline plates, resembling those of *Lepidoderma*. In at least some instances this appearance probably is the result of wetting and drying.

Physarum is the largest genus of the Myxomycetes, numbering over 100 species and, as might be expected, its various species show a wide range of characters. Some fructifications of *Physarum gyrosum* approach *Fuligo* and have been referred to that genus. At the Paris International Botanical Congress in 1954, Locquin suggested that *Fuligo* and *Physarum* be united. This would involve new combinations for nearly all the species of *Physarum*, since *Fuligo* is the older name. Such a step would be unfortunate unless it appeared to be absolutely necessary which, in our opinion, certainly has not been demonstrated. Attempts have been made to divide *Physarum*, as is shown by the synonymy cited above, but in no case can such division be made except on an arbitrary and artificial basis. Some species of *Physarum* very closely approach *Badhamia* and *Craterium*. It may well be that transfer of additional species of *Badhamia* to *Physarum* would make both genera more natural and the former, at least, more homogeneous. *Craterium*, as discussed on p. 69, is sufficiently distinctive to be maintained as a separate genus for the present.

Didymiaceae

Rostafinski, Versuch 12. 1873 (as Tribus).

Capillitium typically limeless, threadlike, purple-brown to pallid, rarely totally absent; peridium usually more or less densely calcareous (limeless in *Diachea*), the lime either in the form of amorphous granules and then aggregated into a shell-like outer layer or peglike protrusions, or imbedded in a cartilaginous wall, or in the form of crystals and then sprinkled over the surface as scattered crystals or platelike scales, or imbedded in cartilaginous walls; spores dark in mass, dark purple-brown to pale violaceous brown by transmitted light.

The Lister and Hagelstein monographs distinguished the Didymiaceae from the Physaraceae by the presence of crystalline as opposed to amorphous lime. Macbride (**1899**) made the distinction between limeless as opposed to limy capillitium, thus agreeing with Rostafinski's original treatment. Neither character is entirely consistent but Macbride's treatment is more workable and it does result in placing *Physarina* and *Diderma* where they seem more at home than in the Physaraceae.

The genus *Diachea* has limy stalks and columellae, but a *Lamprodermalike* peridium and capillitium, thus combining characteristics of the Stemonitales and the Physarales. For this reason the genus has been shifted back and forth between these two orders. (Martin and Alexopoulos, **1969**, relate the early taxonomic history of the genus.) Because of the evidence related on p. 62 and because of absence of lime in the capillitium, *Diachea* is classified in the Didymiaceae, following Farr (**1974**). As stated there, the *Lamprodermalike* characteristics are less weighty, since limeless peridia are found in a few species of *Physarum*, and the capillitium of *Diachea* is fairly similar to that of some didymiaceous species.

Wilczekia Meylan was long maintained as a separate monotypic genus because of the netted capillitium with triangular expansions at the nodes, combined with a peridium sparsely covered with amorphous lime crystals and plasmodic granules. Martin and Alexopoulos (**1969**) alluded to the dubious validity of these characteristics as generic distinctions and recommended its transfer to *Diderma*. This was carried out by Kowalski (**1975**a) who studied Meylan's material and found only granular (not crystalline) lime in *W. evelinae*.

The monotypic genus *Squamuloderma* Kowalski (**1972**) was based on a didymiaceous slime mold lacking capillitium and bearing minute scales on the peridium. The organism was discovered on cow dung in moist chamber and has been grown in laboratory culture by its author and others. The discovery of occasional rudimentary capillitium, coupled with the fact that similar peridial lime aggregations have been reported for several species of *Didymium*, eliminated the basis for maintaining *Squamuloderma* as a separate genus and permitted the transfer of *S. nullifila* to *Didymium* as *D. nullifilum* (Farr, **1982**).

Trabrooksia applanata, type and only species of the genus *Trabrooksia* Keller, is highly suggestive of a limeless form of *Didymium sturgisii*, with which it was compared in the protologue (Keller, **1980**). It has not been grown in culture, nor tested for presence of elemental calcium. Certainly the absence of lime combined with the other traits preclude classifying this slime mold in any other genus. Whether or not the absence of lime is an inherent or environmentally induced character is not quite certain as yet, but so far it has proved stable in numerous collections from various states. The genus is keyed here, at least for purposes of specimen identification.

KEY TO GENERA

a. Lime absent from entire sporophore; capillitium consisting of (or replaced by) vertical hyaline tubules extending from apex to base; sporophores strongly flattened. **Trabrooksia** (see *above*)

aa. Sporophores not combining the above characteristics. b

b. Peridium membranous, iridescent, limeless. c

bb. Peridium membranous to cartilaginous, containing lime. d

c. Peridium tough-membranous with dark granular inclusions, occasionally with a few scattered lime crystals embedded in the thickened base; stalk and columella rarely present, then limeless; capillitium often containing granular nodules. **Leptoderma** (see p. *81)*

cc. Peridium delicate, without granular inclusions, rarely thickened basally; stalk and columella usually present in at least some sporangia of most species, calcareous; capillitium without nodules. **Diachea** (p. *72)*

d. Peridial lime amorphous, granular. e

dd. Peridial lime crystalline, the crystals powdering the surface, united into scales, or forming a continuous crust. f

e. Outer peridium bearing numerous blunt, limy, peglike protuberances. **Physarina** (p. *72)*

ee. Outer peridium without peglike protuberances; middle crystalline layer sometimes present. **Diderma** (p. *73)*

f. Fructification an aethalium. **Mucilago** (p. *73)*

ff. Sporangiate or plasmodiocarpous. g

g. Crystals scattered on peridium or forming a crust, but not united into scales. **Didymium** (p. *74)*

gg. Crystals united into distinct scales, scattered or massed on peridium. **Lepidoderma** (p. *75)*

Diachea

FIGS. 132–137
Plate XIV

Fries, Syst. Orbis Veg. 143. 1825.

Diachaeella Höhnel, Sitz.-Ber. Akad. Wiss. Wien **118**: 436. 1909.

Sporangia globose or cylindric, stipitate or sessile; peridium simple, thin, iridescent, tending to be persistent; columella and stalk, when present, calcareous, rigid, thick, tapering upward; capillitium limeless, delicate, reticulate, the tips of the threads attached to the peridium; spores black or dark brown in mass, pallid, violet-gray, or dark violet-brown by transmitted light.

Type species, *Stemonitis elegans* Trentep. (=*Diachea leucopodia* (Bull.) Rost., Mon. 190. 1874.)

The lime is usually white, but ranges from yellow to orange in two species. It may be crystalline or granular. Usually this structural distinction appears on the species level, but in *D. bulbillosa* the stalks of tropical collections are filled with lime crystals whereas those of temperate-zone specimens invariably have granular lime. Perhaps two species are represented, but there is no other morphological difference between these two groups. Höhnel erected *Diachaeella* on the basis of the crystalline lime.

In the third edition of the Lister monograph (**1925**), three limeless species, *D. cerifera, D. cylindrica,* and *D. caespitosa* were included in the genus. Hagelstein (**1942**) transferred *D. cerifera* to *Elaeomyxa* but retained the other two species in *Diachea*. Macbride and Martin (**1934**), Martin (**1949**), and Martin and Alexopoulos (**1969**) classified *D. cylindrica* and *D. caespitosa* in *Comatricha*, but the latter species was recently shown to have an at least partly calcareous columella at times (Farr, **1979**) and therefore was returned to *Diachea*. Such species as these point toward a possible close relationship between *Diachea* and *Comatricha*.

Of the eight species comprising *Diachea*, only the type species is cosmopolitan and common.

Diachaea is an alternate spelling.

Physarina

FIG. 303
Plate XXXIII

Höhnel, Sitz.-Ber. Akad. Wiss. Wien **118**: 431. 1909.

Sporangia stalked; peridium cartilaginous, appearing single but composed of two closely compacted layers, enclosing amorphous lime granules and bearing on its surface numerous prominent, blunt, subcylindrical, limy pegs, the whole covered with a delicate film, mucous when moist; capillitium limeless, similar to that of *Diderma*; spores purplish black in mass, violet-brown or -gray by transmitted light.

Type species, *P. echinocephala* Höhnel.

As noted by G. Lister in the second and third editions of *The Mycetozoa* (**1911**, **1925**), this genus is very close to *Diderma* and might well be included in it, perhaps as a subgenus, but the two known species are so strikingly different in external appearance from any *Diderma* species that the maintenance of a separate genus is as well justified in this case as in some others.

Physarina consists of two apparently rare species occurring on plant litter. It is known only from the Eastern Hemisphere except for one collection of *P. echinospora* from Mexico (Alexopoulos and Blackwell, **1968**).

Diderma

Pers., Neues Mag. Bot. **1**: 89. 1794.

Leangium Link, Ges. Nat. Freunde Berlin Mag. **3**: 26. 1809.

Polyschismium Corda, Ic. Fung. **5**: 20. 1840.

Chondrioderma Rost., Versuch 3. 1873.

Wilczekia Meylan, Bull. Soc. Vaud. Sci. Nat. **56**: 68. 1925.

FIGS. 304–312 *Plate* XXXIV

FIGS. 313–320 *Plate* XXXV

FIGS. 321–329 *Plate* XXXVI

FIG. 330 *Plate* XXXVII

Fructification usually sporangiate, stalked or sessile, less commonly plasmodiocarpous or pseudoaethaliate; peridium typically double, sometimes apparently single or triple, the outer wall calcareous or cartilaginous, the inner wall membranous, the middle wall, when present, calcareous, amorphous or crystalline; outer wall in calcareous species composed of amorphous lime granules loosely or densely compacted forming a round or smooth shell, the inner layer membranous, delicate, free or closely applied to the outer wall; outer wall in cartilaginous species tough, smooth, often shining, closely applied to the inner or middle wall; columella usually conspicuous, sometimes reduced to a thickened, intrusive, domelike base; capillitium filamentous, branching and anastomosing, limeless; spores dark brown or black in mass, deep purplish brown to pale violaceous brown by transmitted light.

Type species, *Diderma globosum* Pers.

As the name implies, the original emphasis was on the double wall. Persoon listed four species in the original publication. Of these, only *D. globosum* can qualify as the type. *Leangium* was for a time widely recognized as a genus and was later extensively utilized as a subgenus for the species with cartilaginous walls, a usage retained here. Thus, *Diderma* is divided into the two subgenera *Diderma*, based on *D. globosum*, and *Leangium* (Link) Rost., based on *Diderma floriforme* (Bull.) Pers.

The subgenus *Diderma*, with some two dozen species, is about twice as large as subgenus *Leangium*. *Diderma effusum*, *D. testaceum*, *D. hemisphaericum*, and the *D. globosum-crustaceum-spumarioides* complex are undoubtedly the most common representatives of the former subgenus; subgenus *Leangium*, on the whole, is not nearly as common, with *D. floriforme* probably being the most widely distributed member.

It may prove logical in the future to segregate the two species having a calcareous layer in the peridium as at least a separate subgenus. *Polyschismium* would be an available name, even on the generic level (Martin, **1966**).

Mucilago

Micheli ex Batt., Fung. Hist. 76. 1755.

Spumaria Pers., in J. F. Gmelin, Syst. Nat. **2**: 1466. 1791.

FIG. 331 *Plate* XXXVII

Aethalia pulvinate, usually large, consisting of numerous anastomosing tubes filled with spores and capillitium, covered by a dense, calcareous, crystalline cortex, with a delicate membranous inner cortex; capillitium slender, limeless; pseudocapillitium membranous, limy, composed of the walls of the constituent plasmodial strands; spores black in mass.

Type (and only) species, *Mucilago crustacea* Wiggers.

Martin and Alexopoulos (**1969**, pp. *374–375*) discuss the nomenclatural history of this taxon in detail. *Mucilago crustacea*, a common and widely distributed slime mold, is rather variable in spore size, color, and markings, in size of lime crystals, and in consistency of the cortex and interior, but certainly no more so than other common and widely distributed species.

Morgan (**1894**) cited the name as "*Spumaria alba* Bull." and recognized three varieties, var. *didymium* based on *Didymium spumarioides* Fries 1818, not 1829; var. *cornuta*, and var. *mucilago*. G. Lister (**1911**) recognized the varieties *dictyospora* and *solida*. The var. *dictyospora* is described as having an incomplete reticulation on the spores. It was originally reported from Bolivia and later from England, and possibly Minnesota. The var. *solida* has extremely compact aethalia and rather small spores. Jahn (**1924**) raised it to species rank, but here, too, the transition from the variety to the typical form is complete.

Didymium

FIGS. 332–336 *Plate* XXXVII
FIGS. 337–344 *Plate* XXXVIII
FIGS. 345–351 *Plate* XXXIX
FIGS. 352–358 *Plate* XL
FIG. 364 *Plate* XLI

Schrader, Nov. Gen. Pl. 20. 1797.

Lepidodermopsis Höhnel, Sitz.-Ber. Akad. Wiss. Wien **118**: 438. 1909. Not *Lepidodermopsis* Wilczek & Meylan, 1934.

Squamuloderma Kowalski, Mycologia **64**: 1282. 1972.

Fructifications sporangiate or plasmodiocarpous; peridium thin, membranous, rarely cartilaginous, covered with a more or less dense coating of calcareous crystals either scattered loosely over the surface or combined into a crust; columella usually present, sometimes reduced to a thickened calcareous base; capillitium of branching and anastomosing, limeless threads often bearing dark, nodular thickenings (replaced by limy trabeculae in *D. sturgisii*); spores black in mass, violaceous or purplish brown by transmitted light.

Type species, *D. farinaceum* Schrader (=*D. melanospermum* (Pers.) Macbr., N. Am. Slime-Moulds 88. 1899.)

As defined in current usage, *Didymium* is separated from *Diderma* only by the crystalline structure of the lime on the peridium. When the lime crystals are loosely aggregated, as they are in most species, their nature may be inferred with reasonable assurance by examination with a hand lens. When they are united into a crust, only a microscopic mount will reveal their crystalline nature. Crystals, however, may be present under certain conditions in the peridium, stalk, or hypothallus of certain species of *Diderma*, suggesting that the distinction is somewhat arbitrary. However, it is taxonomically workable which, in the present state of our knowledge of the Myxomycetes, justifies its retention.

Of the eight species included in Schrader's original treatment of *Didymium*, most are Didermas. Only *D. farinaceum*, now generally recognized to be the same as the earlier *Physarum melanospermum*, would be included in *Didymium* as it is now defined; hence it must be the type of the genus.

Didymium is usually subdivided into two subgenera, *Didymium* and *Lepidodermopsis* (Höhnel) G. Lister. The latter, based on one species (*D. leoninum*) is of doubtful taxonomic validity or utility (Farr, **1974**).

This sizeable genus of about 35 recognized species contains some universally very common slime molds, such as *Didymium nigripes*, *D. iridis*, *D. squamulosum*, and *D. difforme*. *Didymium ovoideum* is probably more common and widely distributed than indicated in the literature, since it has often been confused with *D. iridis* and *D. megalosporum*.

Lepidoderma

Bary, A. de, in Rost., Versuch 13. 1873.

FIGS. 359–362
Plate XLI

?*Lepidodermopsis* Wilczek & Meylan, Bull. Soc. Vaud. Sci. Nat. **58**: 179. 1934. Not *Lepidodermopsis* Höhnel, 1909.

Sporophores sporangiate or plasmodiocarpous, predominantly sessile; peridium cartilaginous to membranous, covered with crystalline scales, these usually conspicuous but sometimes united into a nearly or quite continuous crust forming a distinct outer wall, sometimes composed of rather loosely compacted crystals; hypothallus membranous to thick and spongy; capillitium as in *Didymium*, typically limeless (in one species with large and expanded nodes which are often vesicular and may enclose clusters of lime crystals); spores black (or nearly so) in mass, violet-brown to dark purplish brown by transmitted light.

Type species, *Didymium tigrinum Schrader* (=*Lepidoderma tigrinum* Rost., in Fuckel, Jahrb. Nass. Ver. Nat. **27–28**: 73. 1873.)

Lepidoderma is closely related to *Didymium*, of which it might well be regarded as a subgenus. The distinctive trait is in the characteristic platelets of crystalline lime. In most collections, these are very striking, but that is not always the case. The variation to nearly continuous crusts of lime on the one hand, and to rather loosely aggregated clusters of lime crystals on the other, may sometimes be seen in a single collection. This variation appears basically comparable to that occurring in *Didymium*, between species having the outer layer (usually of a double peridium) compacted into a crust and those (usually with a single peridium) more or less covered with a powdery or floccose layer of lime.

The type species of *Lepidoderma* is distinctive; the remaining species have been confused but, on the basis of the material available for study, most appear to be discrete.

Lepidoderma, a small genus of six species, is largely confined to mountainous regions, where some species may be locally abundant (Kowalski, **1971**). Only the type species and, possibly, *L. carestianum* appear to be known from lowland localities.

Subclass III

STEMONITOMYCETIDAE

Ross, Mycologia **65**: 483. 1973.

Fructifications mostly sporangiate, sometimes aethaliate or pseudoaethaliate, rarely subplasmodiocarpous, usually covered, at least in early developmental stages, by a peridium which may persist or disappear at maturity; spores germinating by a pore and producing 1–2, rarely more, swarm cells or myxamoebae; sporophore development stemonitoid (epihypothallic), i.e. the hypothallus deposited on the substratum and the protoplasm concentrated above it in one or more masses which develop into sporophores; stalk, when present, secreted internally, continuous with the hypothallus but not the peridium, fibrous or hollow, not stuffed; lime absent (rarely present in trace amounts); gelatinous material occurring in one genus; capillitium present, arising from the columella or the base of the sporophore, sometimes in part from the peridium, or rarely pendent from an apical disk; pseudocapillitium absent; spores mostly dark in mass, occasionally ferruginous (rarely pinkish brown), nearly colorless to yellowish brown or dark purplish brown by transmitted light; assimilative stage, where known, an aphanoplasmodium.

The sole order in this subclass, the Stemonitales, was formerly included, along with all other endosporous orders, in the subclass Myxogastromycetidae. On the basis of its fundamentally different sporophore development and plasmodial structure, the order Stemonitales is now placed in a separate subclass, in accordance with modern concepts discussed on p. *12*.

STEMONITALES

Macbride, N. Am. Slime-Moulds, ed. 2. 122. 1922.

With the characters of the subclass.

Neither the Amaurochaetaceae nor the Lamprodermataceae seem to be sufficiently distinct from the Stemonitaceae to maintain them as separate families and, as Nannenga-Bremekamp (**1967**) pointed out, the family Collodermataceae is equally superfluous.

The family Schenellaceae, although containing only two extremely rare species in one genus, is tentatively accepted here. As indicated by Nannenga-Bremekamp (**1967**), the peculiar capillitial structure sets the genus *Schenella* apart from all other genera of the order.

As discussed on pp. *62* and *71*, *Diachea*, formerly in the Stemonitales, is now classified in the Physarales.

The genus *Symphytocarpus* Ing & Nann.-Brem., judged from the literature and observed specimens, represents a heterogeneous assemblage of *Stemonitis*, *Comatricha*, and *Amaurochaete* species, often in the form of stunted or aberrant developments. Since *Symphytocarpus* has become increasingly adopted in recent years (Eliasson, **1977**b; Nannenga-Bremekamp and Härkönen, **1979**), some discussion is in order.

The type species of *Symphytocarpus* is *S. flaccidus* (A. Lister [erroneously cited as "Morgan"]) Ing & Nann.-Brem., based on *Stemonitis splendens* var. *flaccida* A. Lister. In their protologue, Ing and Nannenga-Bremekamp (**1967**) list *Comatricha flaccida* (A. Lister) Morgan and *Amaurochaete ferruginea* Macbr. & Martin as additional synonyms. Later on, however, Nannenga-Bremekamp (**1974**) recognized the dissimilarity between *Comatricha flaccida* and *Amaurochaete ferruginea*, a fact earlier brought out by Martin and Alexopoulos (**1969**), who had studied some of Morgan's specimens. In my (Farr's) opinion, one of these Morgan collections represents densely aggregated but clearly recognizable sporangia of *Stemonitis splendens*, while the other is more confluent and irregular. Neither, however, could be mistaken for *Amaurochaete*. Morgan's description (**1894**), incidentally, takes into account the variations in columella development.

I have not seen the type specimen of *Symphytocarpus flaccidus*, but an authentic collection (*Nannenga-Bremekamp 9031*) in BPI is *Amaurochaete ferruginea*. The illustrations of the species, on the other hand (Ing and Nannenga-Bremekamp, **1967**, *fig. 1*) certainly do not depict *A. ferruginea*, but clearly suggest *Stemonitis splendens* var. *flaccida*. A study of many herbarium collections labelled as the latter, and as the misunderstood *Comatricha flaccida* (and the comments accompanying the specimens), indicates considerable confusion regarding the concepts of these taxa. A comparison of them with type and other collections of *Amaurochaete ferruginea* shows that all of the material is either referable to the latter species or represents an intergrading series of atypical to fairly typical examples of *Stemonitis splendens*. Some specimens show a range from agglutinated to discrete sporangia, as well as from total absence to partial presence of surface net, within the same colony. The capillitium and spores of these sporangia, if well-developed, are usually fairly typical for *Stemonitis splendens*. Currie, as early as 1919 (p. 299), provides an entirely adequate description of *Stemonitis splendens* var. *flaccida*, and that author's comments (just preceding the description) on "confluent types" of *Stemonitis* spp. in general, certainly apply today. The frequent evidence in the anomalous fruitings (as well as in a number of specimens of *Stemonitis fusca* var. *confluens*) of ?insect activity suggests that disturbance or feeding by minute animals or, perhaps, heavy rain on a mounded plasmodium starting to differentiate, possibly may disrupt formation of normal columellae and other aspects of sporangial development, such as surface net formation. Eliasson (**1977**a) found these factors to affect fruiting body morphology in *Amaurochaete*. The varieties *flaccida* and *confluens* described for at least three species of *Stemonitis* may represent such developments (*Stemonitis confluens* Cooke & Ellis, however, appears correctly interpreted as a distinct species).

These observations indicate that *Amaurochaete ferruginea* is not conspecific with *Stemonitis splendens* var. *flaccida*, as postulated by Hagelstein (**1944**), Ing and Nannenga-Bremekamp (**1967**), and Nannenga-Bremekamp (**1974**). Furthermore, since other species of the genus *Amaurochaete* also lack surface nets and often show no evidence of a cortex, there does not seem to be a critical need at present to displace *A. ferruginea* into a separate genus. *Symphytocarpus* is said to differ from *Amaurochaete* by lacking a cortex and being pseudoaethaliate rather than aethaliate. It is true that many fructifications of *Amaurochaete* display a more or less definitely aethaliate habit but, in the majority of specimens of several species, the cortex seems to be lacking and whether it was present during developmental stages is not known. In at least one species, *A. comata*, which has been cultured spore to spore (Farr, **1982**), no cortex was observed at any stage. Martin (**1961**b) raised the possibility that "the fructifications of all species of *Amaurochaete* should be classed as pseudoaethalia." Those species with a cortex, however, probably should be considered as aethaliate. The possibility of genetic influence on, or regulation of, such factors as columella development and surface net should not be ruled out. The fact that these same aberrations occur in several species of *Stemonitis*, as indicated above, would tend to support the theory that these varieties represent irregularities rather than distinct taxa. (It may be noteworthy in this connection that varieties *flaccida* and *confluens* of *Stemonitis fusca* are both listed as synonyms of *Symphytocarpus amaurochaetoides*.)

KEY TO FAMILIES

a. Capillitium consisting of numerous tortuous threads more or less coiled into columnar strands, each strand representing one sporangium, attached at the cuplike peridial base and the peridial lid; sporophores pseudoaethaliate, covered by an evanescent, membranous cortex; peridia fugacious except for a basal cup and apical lid. **Schenellaceae** (p. *78*)

aa. Capillitium not as in A, usually more or less reticulate (occasionally greatly reduced), arising from the columella, the sporangial base, or (in one genus) pendent from an apical disk. **Stemonitaceae** (p. *79*)

Schenellaceae

Nannenga-Bremekamp, Proc. Nederl. Akad. Wet., ser. C, **70**: 203. 1967.

Sporophores pseudoaethaliate, covered by a membranous, continuous cortex; sporangia columnar, erect, each with a cuplike constricted base and persistent lid which is firmly attached to the cortex; peridium evanescent except for the basal cup and apical lid; columella absent; capillitium consisting of sinuous, unbranched or sparsely branched threads attached basally and apically and more or less coiled into columnar strands, each strand representing a sporangial unit.

With one genus, *Schenella*.

Macbride, in proposing the genus, gave an excellent description of the type species, but he was not entirely convinced that it belonged to the Myxomycetes and did not mention it in the second edition of his monograph (**1922**) nor did G. Lister in the third edition of *The Mycetozoa* (**1925**). Macbride and Martin (**1934**) did recognize it, but both Hagelstein (**1944**) and Martin (**1949**) made only casual reference to it as doubtful. The finding of a second species convinced Martin (**1961**b) that *Schenella* is a valid myxomycete genus properly placed in the Stemonitales. As Martin observed, the genus shows some similarities with the Dianemaceae (Trichiales) and the Stemonitaceae. He interpreted the sporophore as a pseudoaethalium somewhat comparable with *Dictydiaethalium* and *Tubifera* (Liceales), but also similar to the fructification of *Amaurochaete* (which he likewise regarded as a pseudoaethalium). In placing *Schenella* in the Stemonitaceae he stressed the dark color of capillitium and spores (of *S. simplex*). *Schenella microspora*, described later, is ferruginous, however, and thus similar in color to *Tubifera*.

Schenella

Macbr., Mycologia **3**: 39. 1911.

FIGS. 138–139
Plate XIV

With the characters of the family.

Type species, *Schenella simplex* Macbr.

Although the two species of *Schenella* are so far known only from their type specimens, the latter are ample and most of their diagnostic characters well preserved. Only fragments of the cortex are left in the type of *S. simplex*, but the capillitial bundles are distinct. The type collection of *S. microspora* at this time is a mass of capillitium and spores. While additional mature collections certainly would be welcomed, they probably would not shed any additional light on what is already known about sporophore structure. What is needed, of course, are developmental studies from fresh material to elucidate plasmodium structure and sporophore differentiation, in order to clarify the taxonomic position of this enigmatic genus.

Stemonitaceae

Rost. Versuch. 6. 1873 (as Tribus).

Fructifications sporangiate to aethaliate or pseudoaethaliate; peridium membranous, persistent or fugacious, in one genus gelatinous; columella present in sporangiate forms (except in *Diacheopsis*, *Colloderma*, and sometimes *Leptoderma*); capillitium threadlike, branching and usually anastomosing, arising from the columella, from the base of the sporophore, or sometimes in part from the peridium or, in one genus pendent from an apical disk; hypothallus membranous, sometimes common to a group of sporangia; spores black, purplish brown, or ferruginous in mass.

Nannenga-Bremekamp (**1967**) and Ing and Nannenga-Bremekamp (**1967**) proposed radical emendation of the Stemonitaceae. Placing primary emphasis on stalk structure, Nannenga-Bremekamp removed those species of *Stemonitis* which have fibrous stalks to *Stemonitopsis*, a new subgenus of *Comatricha*. Two new genera were erected: *Collaria* Nann.-Brem., which includes three species transferred from *Comatricha*, and *Symphytocarpus* Ing & Nann.-Brem., comprising some species formerly referred to *Stemonitis* and *Amaurochaete*, together with several additional taxa. The genus *Symphytocarpus* is discussed on p. *77*. These authors also recognized Hertel's genera *Paradiachea* and *Paradiacheopsis* (based on *Diachea cylindrica* and *Paradiacheopsis curitibana*, respectively). The former genus comprises three sessile species withdrawn from *Comatricha* and *Diachea* (with a new species added later), whereas the latter is restricted to those *Comatricha* species that have simple capillitium and stalks with fibrous bases.

On the other hand, Nannenga-Bremekamp later (**1974**) proposed a merger of the three genera *Diacheopsis*, *Colloderma*, and *Leptoderma*. As indicated on p. *81*, such a union at this time would be premature.

KEY TO GENERA

a. Fructification an aethalium or pseudoaethalium. — b

aa. Fructification sporangiate, the sporangia scattered, clustered, or rarely massed into a pseudoaethalium. — c

b. Capillitium a network of more or less horizontal, branching threads, the tips of the branches united by many chambered vesicles. — **Brefeldia** (p. *80*)

bb. Capillitium dendroid, the main stalks arising from the base, the tips of the branches free. — **Amaurochaete** (p. *80*)

c. Wax present in one or more parts of the sporophore. — **Elaeomyxa*** (see p. *62*)

cc. Wax not secreted. — d

d. Columella usually lacking or merely in the form of a dome-shaped basal

---

**Elaeomyxa* has been moved to the Physarales, but is also keyed here because of its traditional classification in the Stemonitales.

thickening; sporangia sessile on broad or constricted bases which rarely are contracted into short, thick stalks. e

dd. Columella present (lacking only in atypical developments); sporangia mostly stalked, rarely sessile and then with a distinct columella from which the capillitium arises. g

e. Outer layer of peridium gelatinous when wet. **Colloderma** (p. *81)*

ee. Outer layer of peridium not gelatinous when wet. f

f. Peridium membranous above, thickened with granular deposits below and sometimes including scalelike masses of lime crystals in lower part and hypothallus. **Leptoderma** (p. *81)*

ff. Peridium membranous, hyaline, iridescent throughout; lime crystals lacking. **Diacheopsis** (p. *82)*

g. Columella enlarged at apex into a cupulate disk from which the capillitium depends. **Enerthenema** (p. *82)*

gg. Columella not bearing a cupulate apical disk; capillitium usually arising from the entire columella or from the sporangial base. h

h. Peridium evanescent, but typically replaced by a surface net developed under the periphery, remaining after the peridium has been shed, and united with the capillitium. **Stemonitis** (p. *82)*

hh. Peridium persistent or, if evanescent, without surface net; capillitium scanty to abundant, without surface net and often with many terminal branchlets. i

i. Stalk typically translucent, hollow, often yellow at base. **Macbrideola** (p. *83)*

ii. Stalk dark, opaque, not hollow. j

j. Peridium tough, metallic, shining, tending to be long persistent as a whole or at least basally. **Lamproderma** (p. *84)*

jj. Peridium usually early evanescent or, if persistent, then membranous and delicate. **Comatricha** (p. *84)*

Brefeldia

FIG. 124
Plate XIII

Rost., Versuch. 8. 1873.

Fructification a large, pulvinate aethalium with a continuous cortex, arising from a broadly expanded hypothallus, broken up internally by irregular walls and numerous flattened, columellalike projections, the latter giving rise to the threadlike, netted capillitium, which bears inflated, multicellular vesicles at the nodes; spores black in mass.

Type (and only) species, *Reticularia maxima* Fr. (=*Brefeldia maxima* (Fr.) Rost. in Fuckel, Jahrb. Nass. Ver. Nat. **27–28**: 70. 1873.)

Brefeldia maxima has a widespread distribution in America and Europe, but is uncommon (at least in the United States). It is one of the largest known slime molds, whose fructifications may reach 30 cm in the longer dimension (although usually they are much shorter). The suggestion of the basic presence of numerous sporangia is very apparent in some collections, and the numerous columellalike stalks arising from the base of the aethalium further suggest this. The characteristic vesicular structures may arise at the junction of the confluent sporangia somewhat as do the peridial disks in *Stemonitis confluens*. Nevertheless the picture is that of a true aethalium.

Amaurochaete

FIGS. 125–129
Plate XIII

Rost., Versuch. 8. 1873.

Matruchotia Skup., Bull. Acad. Pol. 1924: 396. 1927. Not *Matruchotia* Boul., 1893.

Matruchotiella Skup. ex G. Lister, Mycet. ed. 3. 165. 1925. Not *Matruchotiella* Grigor., 1924.

Fructification aethaliate or pseudoaethaliate, depressed-pulvinate; cortex, when present, evanescent, leaving after its disappearance a mass of irregular stalks and branches seated on a common, membranous hypothallus; spores black or brown.

Type species, *Lycogala atrum* Alb. & Schw. (=*Amaurochaete atra* (Alb. & Schw.) Rost. Mon. 211. 1874.)

The suggestion that the fructifications are composed of densely massed and agglutinated sporangia is very strong in some collections, which is the reason certain species have been regarded as representing aberrant sporophores of *Stemonitis*. That is particularly true of *Amaurochaete ferruginea*. However, the complete lack of a surface net in all specimens examined, and especially the dendroid character of the erect pillars regarded as columellae in such species make this doubtful, especially when the spore characters are compared with those of the species of *Stemonitis* to which they are referred. It is more likely that some of the sporophores might be regarded as pseudoaethalia rather than aethalia, but only careful study of younger stages than are available, and particularly of developmental stages, could clarify this point.

All of the five species of this rather uncommon genus are known from North America, but *Amaurochaete atra* appears to be the most frequently collected.

Colloderma

G. Lister, J. Bot. **48**: 312. 1910.

FIGS. 122–123
Plate XIII

Sporangia sessile; peridium double, the outer layer more or less gelatinous when moist, drying horny, the inner layer membranous; columella lacking; capillitium slender, hyaline to dark, radiating from the base, branching and anastomosing to form a net; spores dark.

Type species, *Didymium oculatum* Lipp. (=*Colloderma oculatum* (Lipp.) G. Lister, J. Bot. **48**: 312. 1910.)

Nannenga-Bremekamp (**1967**) stated that an outer gelatinous wall occurs in other genera and is not, in such cases, regarded as of sufficient importance to justify a separate genus, much less a family; hence she included *Colloderma* in the Stemonitaceae, and her treatment has been accepted by other modern monographers. Within the Stemonitaceae the genus is unique in having a double peridium of which the outer, more or less gelatinous layer softens and sometimes swells upon wetting.

In habit and capillitial structure *Colloderma* appears close to *Diacheopsis* and further studies may well support the previously suggested merger of the two genera. (This, of course, would require an expanded genus concept, as well as nomenclatural recombinations in *Colloderma*, the older name.) As in the case of *Leptoderma*, developmental studies are needed to clarify the taxonomic position of *Colloderma*.

Of the two species described in *Colloderma*, only the type species is known from America. Even this seems to be rare (or, being very inconspicuous in the dry state, perhaps overlooked). A Mexican specimen appearing intermediate between the two species was reported by Farr (**1976**b) and described in detail by Braun and Keller (**1976**).

Leptoderma

G. Lister, J. Bot. **51**: 1. 1913.

FIG. 363
Plate XLI

Sporangia sessile (rarely with a short, thick stalk); peridium tough-membranous, thickened with dark, granular inclusions at the base, where small scalelike aggregations of crystalline lime may be imbedded; capillitium dark, often with pale extremities, netted, arising from the sporangial base; spores dark in mass.

Type (and only) species, *L. iridescens* G. Lister.

This is a most peculiar genus, based on a single species which suggested to its author a nearly limeless phase of a *Lepidoderma*. Syntype material and other collections of *L. iridescens*, however, do not fit into *Lepidoderma* or any other genus. *Leptoderma* thus should be retained provisionally, but its taxonomic position is enigmatic. Superficially, the subglobose to pulvinate, gray to iridescent, limeless sporangia show considerable resemblance not only to *Diacheopsis*, but also to limeless fructifications of certain members of the Physarales which, according to Kowalski (**1975**b), mimic the latter genus. A merger of *Leptoderma* and *Diacheopsis* has been suggested more than once (Martin and Alexopoulos, **1969**; Nannenga-Bremekamp, **1974**), but would be premature at this point in our knowledge. Although the capillitium certainly brings to mind that of a *Lamproderma* (G. Lister, **1913; 1925**; Hagelstein, **1944**; and others), other characteristics seem to rule out a close relationship with the Stemonitales. The dark debris and the sporadic occurrence of lime crystals, as well as the occasional presence of a debris-filled stalk (G. Lister, **1913**; and subsequent authors) hint at a subhypothallic mode of development. Thus the taxonomic status of *Leptoderma* will probably not be clarified until developmental stages can be observed. The genus is retained in the Stemonitaceae as a concession to tradition, but keyed also in the Didymiaceae in order to facilitate specimen identification.

Leptoderma iridescens is known on this continent with certainty only from the mountainous regions of the West, the best representative specimens being Kowalski's collections from Colorado and California. A collection by Thaxter from Maine tentatively referred to this species and now deposited in the New York Botanical Garden (*Hagelstein 7942*), may represent the taxon, but is devoid of lime and capillitium. Another Maine specimen determined by Hagelstein (Farlow Herbarium 2980) appears to be *Physarum nudum*.

Diacheopsis

Meylan, Bull. Soc. Vaud. Sci. Nat. **57**: 149. 1930.

Fructifications sessile or borne on weak, strandlike stalks, sporangiate to plasmodiocarpous, scattered to clustered, subglobose to pulvinate; peridium usually persistent, membranous, delicate, translucent, iridescent; columella none; capillitium variable, arising from the sporangial base, branching and anastomosing, forming a loose to dense net, colorless to dark purple-brown, sometimes with nodular expansions; spores dark.

Type species, *D. metallica* Meylan.

Differing from *Lamproderma* in the lack of a columella and in the origin of the capillitium. Kowalski (**1975**b), on whose monograph the present description is partly based, recognized six species in this uncommon montane genus, of which only *D. metallica* is often collected.

Enerthenema

FIGS. 140–142
Plate XV

Bowman, Trans. Linn. Soc. **16**: 152. 1830.

Ancyrophorus Raunk., Bot. Tidsskr. **17**: 92. 1888.

Sporangia stipitate, the stipe continued as a columella to the top of the sporangium and there expanding into a shallow cup or disk from which the capillitium depends; peridium fugacious; spores dark.

Type species, *E. elegans* Bowman (=*E. papillatum* (Pers.) Rost., Mon. App. 28. 1876.)

Of the three species comprising this well-defined genus, *Enerthenema papillatum* is by far the most common and has a wide distribution. *Enerthenema melanospermum* Macbr. & Martin is a mountain species of the western United States.

Stemonitis

FIGS. 143–146
Plate XV

Roth, Mag. Bot. Römer & Usteri **1** (2): 25. 1787 (*Nomen conserv.*).

Symphytocarpus Ing & Nannenga-Bremekamp, Proc. Nederl. Akad. Wet., ser. C, **60**: 218, 1967.

Stemonitopsis (Nann.-Brem.) Nann.-Brem., Nederl. Myxomyceten. 203. 1974.

FIGS. 147–153 *Plate* XVI

FIGS. 154–156 *Plate* XVII

Sporangia cylindrical, stalked, gregarious or densely clustered, rarely forming a pseudoaethalium; hypothallus membranous, usually well developed, often common to a cluster; stalk extending into the sporangium as a columella; capillitium arising from the entire length of the columella, branching repeatedly, the final branches united with a usually more or less persistent surface net developed under the peridium; peridium fugacious at maturity, leaving the spore mass enclosed by the surface net; spores black, fuscous, brown, or ferruginous in mass, violaceous-brown to nearly colorless by transmitted light.

Type species, *Stemonitis fusca* Roth.

Stemonitis is distinguished from *Comatricha* by the presence of a surface net, developed just under the peridium and independently from the capillitium arising from the columella, and fusing with the ultimate branches of the capillitium. The net does not appear to be a part of the peridium remaining after the interstices have dropped away, as in *Cribraria*. Species of *Comatricha* with dense capillitium, such as *C. typhoides* and *C. pulchella*, may appear to have a surface net, but careful observation indicates that is not the case. Furthermore, those species of *Comatricha* whose sporangial development has been studied differ in their developmental pattern (particularly of the capillitium) from those of *Stemonitis* species (Ross, **1958**b; Goodwin, **1961**). Nevertheless, the distinction is not always clear and the newer classifications, while not yet satisfactory, may suggest the way to future clarification.

Many of the species are difficult to define and hence difficult to key. An important key character, frequently used, is the presence or absence of reticulations on the spores. Sometimes these are clear, as in *Stemonitis trechispora* and many collections of *S. fusca*. More often they are faint and hard to see except with the use of an oil immersion objective. However, after some experience, they may usually be detected under a high dry objective focussed carefully on the spore surface and with illumination properly adjusted.

Nannenga-Bremekamp (**1967**) restricted *Stemonitis* to species with cylindrical sporangia, well-marked surface nets, and smooth, hollow stalks, with particular emphasis upon the character of the stalks. This was more fully developed by Ing and Nannenga-Bremekamp (**1967**), where most of the excluded species were referred to their new genus *Symphytocarpus* (see p. 77).

Stemonitis is a large, ubiquitous genus of about twenty species, of which at least three are common and cosmopolitan, and several others appear to be nearly so. All species are macroscopic, with sporangia ranging from slightly under 2 mm up to 20 mm or more in length. In the field the genus is easily recognized except for a few species that may be mistaken for *Comatricha*.

Stemonites and *Stemionitis* are alternative spellings found in the literature.

Macbrideola

H. C. Gilbert, Univ. Iowa Stud. Nat. Hist. **16**: 144. 1934, emend. Alexop., Mycologia **59**: 112. 1967.

Paradiacheopsis Hertel, Dusenia **5**: 191. 1954, *p. p.*

FIG. 161 *Plate* XVII

FIG. 170 *Plate* XVIII

FIG. 179 *Plate* XIX

FIGS. 182–183 *Plate* XX

Sporangia minute, stipitate; peridium membranous, translucent, early evanescent or persistent; stipe hollow, tubular, typically translucent, often with a yellow base, extending into the sporangium as a columella; capillitium present or absent, when present varying from a few short branches of the columella to a very open globose net, usually rising from the tip of the columella, occasionally from the side; spores dark in mass, pallid, brown, or violet-brown by transmitted light.

Type species, *M. scintillans* H. C. Gilb.

As emended (Alexopoulos, **1967**), *Macbrideola* permits the inclusion of several small species with scanty capillitium and hollow stalks, formerly included in *Comatricha*. Such modification makes both genera more homogeneous.

Because of the small size of the sporangia, the five species of this genus are known predominantly from moist-chamber cultures. This may account in part for the widely scattered localities comprising its reported range and its relatively poor representation in herbaria.

Lamproderma

FIGS. 186–190 *Plate* XX

FIGS. 191–199 *Plate* XXI

Rost., Versuch 7. 1873.

Sporophores sporangiate, rarely plasmodiocarpous; sporangia globose or ellipsoid to subcylindrical or fusoid, stalked or sessile on a constricted base, rarely pulvinate; peridium thin- to tough-membranous, persistent to partly (rarely entirely) evanescent, usually strongly iridescent but sometimes only lustrous or dull-metallic; columella usually conspicuous, cylindric or clavate, attaining from one-third to two-thirds the height of the sporangial cavity, rarely shorter, occasionally reaching the tip; capillitium arising mainly from the tip of the columella, densely reticulate, usually dark, the branches thinner and often paler as they approach the periphery; spores dark in mass.

Type species, *Physarum columbinum* Pers. (=*Lamproderma columbinum* (Pers.) Rost. in Fuckel, Jahrb. Nass. Ver. Nat. **27–28**: 69. 1873.)

Lamproderma appears to be closely related to *Comatricha* (Martin and Alexopoulos, **1969**; Kowalski, **1970**; Farr, **1976**b). It is distinguished from those Comatrichas with abundant capillitium by the usually much more persistent peridium and by the tendency of the capillitium to originate at or near the tip of the columella. Old, weathered specimens of *Lamproderma*, in which the peridium has completely disappeared, can be separated from *Comatricha* only by the aspect of the capillitium (sometimes also by the presence of a basal ring or collar representing peridial remnants around the stipe) and by the spores, which are often distinctive.

The numerous species and subspecific taxa recognized by various authors, especially Meylan (**1932**), emphasize the difficulties of determining the limits of species in *Lamproderma*. At present it seems best to interpret the evidence as indicating a wide capacity for variation in most of the species and to enlarge concepts of the species to conform with that interpretation, as was done in the early study of Dennison (**1945**). More recently the genus was monographed by Kowalski (**1970**) who recognized 21 species, the great majority of which are montane. *Lamproderma* as a whole appears to be one of the more clearly delimited, homogeneous genera. *Lamproderma arcyrionema* and *L. scintillans* are typical representatives and probably the most widely known species of the genus. The former species was transferred by Nannenga-Bremekamp (**1967**) to a new genus, *Collaria*.

Comatricha

FIGS. 157–160, 162–164 *Plate* XVII

FIGS. 165–169, 171–173 *Plate* XVIII

FIGS. 174–178, 180–181 *Plate* XIX

Preuss, Linnaea **24**: 140. 1851.

Rostafinskia Racib., Rozp. Akad. Umiej. **12**: 77. 1884. Not *Rostafinskia* Speg., 1880.

Raciborskia A. Berl. in Sacc. Syll. Fung. **7**: 400. 1888.

Paradiacheopsis Hertel, Dusenia **5**: 191. 1954, *p. p.*

Comatrichoides Hertel, Dusenia **7**: 347. 1956. Invalid.

Paradiachea Hertel, Dusenia **7**: 349. 1956.

Collaria Nann.-Brem., Proc. Nederl. Akad. Wet., ser. C, **70**: 208. 1967.

Sporangia cylindric to globose, scattered to gregarious to densely crowded, minute to several cm long; columella typically present, usually reaching nearly to the apex of the sporangium, rarely lacking, giving rise to numerous branches which subdivide and often anastomose to form a capillitial net, the ultimate branchlets usually free; peridium free, usually evanescent, sometimes persistent; spores black, purple, or ferruginous in mass, violaceous brown to pallid by transmitted light.

Type species, *Stemonitis obtusata* Fr. (=*Comatricha nigra* (Pers.) Schroeter, Krypt.-Fl. Schles. **3** (1): 118. 1885.)

Comatricha is closely related to *Stemonitis*, from which it is separated by the lack of the surface net characteristic of the latter genus. Even in such species as *Comatricha typhoides* and *C. laxa*, where the capillitium is extensively branched and anastomosed near the surface, the lack of a surface net is apparent, whereas it is nearly always present in *Stemonitis* sporangia, at least in the lower portion, and to a greater extent than the descriptions imply in those species in which it is said to be lacking above. In *S. hyperopta*, for example, there are often traces of delicate net in the upper part of the sporangium, although the net falls away more quickly than in other species.

Ross (**1958**b) showed that, in three species of *Stemonitis*, the capillitium developed both from the columella and from areas within the sporogenous mass and that the net developed just under the surface from loci within the peripheral protoplasm, while in *Comatricha typhoides* the entire capillitial system arose from the columella. Goodwin (**1961**) studied three additional species of *Comatricha* and found that they developed as described by Ross for *C. typhoides*. This suggests that the presence or absence of a net may be of fundamental significance.

Alexopoulos (**1967**) stressed the importance of stalk characters in these genera; stalks of *Stemonitis* and *Macbrideola* typically are hollow, tubular, and homogeneous, while those of *Comatricha* are filled with an interlaced mass of threadlike strands. Partly on that basis he transferred three of the minute species of *Comatricha* to *Macbrideola*.

Hertel (**1956**) earlier suggested extensive revision of *Comatricha*, proposing several new genera. His work was reviewed and amplified by Nannenga-Bremekamp (**1967**). The latter author recognized *Paradiachea* Hertel essentially as it was originally proposed, to include those species with a persistent peridium. In view of the quantitative variation of this character among and even within species (*Comatricha typhoides*, for example), this separation appears superfluous. Nannenga-Bremekamp (**1967, 1974**) also accepted and substantially modified *Paradiacheopsis* Hertel, transferring several additional species to that genus, and distributed the residual species of *Comatricha* among four subgenera: *Comatricha*, *Laxaria* Nann.-Brem., *Sinuaria* Nann.-Brem., and *Stemonitopsis* Nann.-Brem. The last-named was raised to generic rank in her book (**1974**) and typified by *Stemonitis hyperopta*.

Although *Comatricha* in the classical sense obviously represents a large, heterogeneous assemblage of species, such drastic revision seems unwarranted. Some of the characteristics used in segregating the aforementioned genera (degree of ramification in the capillitium, amount of persistent peridium, etc.) may intergrade among species or vary even within a population. Perhaps for the sake of uniformity *Comatricha* should be subdivided, at least into subgenera, but the difficulty is where to draw the lines.

Comatricha is a large, universally distributed genus. In the classical sense adopted here, it consists of slightly over two dozen species, of which *C. typhoides*, *C. pulchella*, and *C. nigra* are probably the most prevalent. In the tropics, *Comatricha longa*, a species of considerable superficial resemblance to *Stemonitis splendens*, likewise is common.

XIV. Bibliography

Adanson, M. 1763. Familles des plantes. Vols. **1** and **2**. Paris.

Ainsworth, G. C. 1973. Introduction and keys to higher taxa. In Ainsworth, G. C., F. K. Sparrow, and A. S. Sussman. The Fungi IVA.: 1–7. Academic Press, New York.

Aldrich, H. C. 1967. The ultrastructure of meiosis in three species of *Physarum*. Mycologia **59**: 127–148.

______. 1968. The development of flagella in swarm cells of the Myxomycete *Physarum flavicomum*. J. Gen. Microbiol. **50**: 217–222.

______, and M. M. Blackwell. 1976. Resistant structures in the Myxomycetes. In Weber, D. J., and W. M. Hess (eds.). The fungal spore, form and function. John Wiley & Sons, Inc. New York.

______, and G. Carroll. 1971. Synaptonemal complexes and meiosis in *Didymium iridis*; a reinvestigation. Mycologia **63**: 308–316.

______, and J. W. Daniel (eds.). 1982. Cell Biology of *Physarum* and *Didymium*. Two vols. Academic Press, New York.

______, and C. W. Mims. 1970. Synaptonemal complexes and meiosis in Myxomycetes. Amer. J. Bot. **57**: 935–941.

Alexopoulos, C. J. 1958. Three new species of Myxomycetes from Greece. Mycologia **50**: 50–56.

______. 1960a. Morphology and laboratory cultivation of *Echinostelium minutum*. Amer. J. Bot. **47**: 37–43.

______. 1960b. Gross morphology of the plasmodium and its possible significance in the relationships among the Myxomycetes. Mycologia **52**: 1–20.

______. 1961. A new species of *Echinostelium* from Greece. Amer. Midl. Nat. **66**: 391–394.

______. 1962. Introductory mycology. Ed. 2. John Wiley & Sons, New York.
______. 1964a. The rapid sporulation of some Myxomycetes in moist chamber culture. Southwestern Nat. **9**: 155–159.
______. 1964b. The white form of *Physarella oblonga*. Mycologia **56**: 550–554.
______. 1967. Taxonomic studies in the Myxomycetes 1. The genus *Macbrideola*. Mycologia **59**: 103–116.
______. 1969. The experimental approach to the taxonomy of the Myxomycetes. Mycologia **61**: 219–239.
______. 1973. Myxomycota. Myxomycetes. In Ainsworth, G. C., A. S. Sussman, and F. K. Sparrow (eds.). The Fungi IVB: 39–60. Academic Press, New York.
______. 1982. Morphology, taxonomy, and phylogeny. Chapter 1 (pp. 3–23) in Aldrich, H. C., and J. W. Daniel (l.c., vol. 1).
______, and M. M. Blackwell. 1968. Taxonomic studies in the Myxomycetes II. *Physarina*. J. Elisha Mitchell Soc. **84**: 48–51.
______, and T. E. Brooks. 1971. Taxonomic studies in the Myxomycetes III. Clastodermataceae: a new family of the Echinosteliales. Mycologia **63**: 925–928.
______, and C. W. Mims. 1979. Introductory mycology. Ed. 3. John Wiley & Sons, New York.
Anderson, J. D. 1962. Potassium loss during galvanotaxis of slime mold. J. Gen. Physiol. **45**: 567–574.
______. 1964. Regional differences in ion concentration in migrating plasmodia. In Allen, P. J., and N. Kamiya (eds.). Primitive motile systems in cell biology, pp. 125–136. Academic Press, New York.
Arambarri, A. M. 1975. Myxophyta, Myxomycetes. Flora cript. Tierra del Fuego **2**. Buenos Aires.
Baker, G. 1933. A comparative morphological study of the myxomycete fructification. Univ. Iowa Stud. Nat. Hist. **14**(8): 1–35.
Bary, A. de. 1858. Über die Myxomyceten. Bot. Zeit. **16**: 357–358; 361–364; 365–369.
______. 1859. Die Mycetozoen. Ein Beitrag zur Kenntnis der niedersten Thiere. Zeits. f. wiss. Zool. **10**: 88–175.
______. 1864. Die Mycetozoen (Schleimpilze). Ein Beitrag zur Kenntnis der niedersten Organismen. Engelmann, Leipzig.
______. 1884. Vergleichende Morphologie und Biologie der Pilze, Mycetozoen, und Bacterien. Engelmann, Leipzig.
Batsch, A. J. G. C. 1783–1789. Elenchus fungorum. Halle.
Berlese, A. N. 1888. Myxomyceteae Wallr. In Saccardo, P. A., Syll. Fung. **7**: 323–453.
Bessey, E. A. 1950. Morphology and taxonomy of fungi. The Blakiston Co., Philadelphia.
Bisby, G. R. 1914. Some observations on the formation of the capillitium and the development of *Physarella mirabilis* Peck and *Stemonitis fusca* Roth. Amer. J. Bot. **1**: 274–288.
Bjørnekaer, K., and A. B. Klinge. 1963. Die dänischen Schleimpilze. Myxomycetes daniae. Friesia **7**: 149a–280.
Blackwell, M. M. 1974. A study of sporophore development in the Myxomycete *Protophysarum phloiogenum*. Arch. Microbiol. **99**: 331–344.
______, and C. J. Alexopoulos. 1975. Taxonomic studies in the Myxomycetes IV. *Protophysarum phloiogenum*, a new genus and species of Physaraceae. Mycologia **67**: 32–37.
______, and R. L. Gilbertson. 1980. *Didymium eremophilum*: a new Myxomycete from the Sonoran desert. Mycologia **72**: 791–797.
______, T. G. Laman, and R. Gilbertson. 1982. Spore dispersal in *Fuligo septica* (Myxomycetes) by lathridiid beetles. Mycotaxon **14**: 58–60.
Bonner, J. T. 1967. The cellular slime molds. Ed. 2. Princeton Univ. Press, Princeton.
Braun, K. L., and H. W. Keller. 1976. Myxomycetes of Mexico I. Mycotaxon **3**: 297–317.
Brefeld, O. 1869. *Dictyostelium mucoroides*. Ein neuer Organismus aus der Verwandschaft der Myxomyceten. Abh. Senckenberg. Naturf. Ges. Frankfort **7**: 85–107.
Brown, R. M., Jr., D. A. Larson, and H. C. Bold. 1964. Airborne algae: their abundance and heterogeneity. Science **143**: 583–585.
Bulliard, J. B. F. 1791. Histoire des champignons de la France. I.
Buxbaum, J. C. 1721. Enum. Plant. in agro Halensi. Halle.

Camp, W. G. 1937. The structure and activities of myxomycete plasmodia. Bull. Torrey Bot. Club **64**: 307–335.

Carlile, M. J. 1970. Nutrition and chemotaxis in the myxomycete *Physarum polycephalum*: the effect of carbohydrates on the plasmodium. J. Gen. Microbiol. **63**: 221–226.

______. 1972. The lethal interaction following plasmodial fusion between two strains of the Myxomycete *Physarum polycephalum*. J. Gen. Microbiol. **71**: 581–590.

______. 1973. Cell fusion and somatic incompatibility in Myxomycetes. Ber. Deuts. Bot. Ges. **86**: 123–139.

______. 1974. Incompatibility in the Myxomycete *Badhamia utricularis*. Trans. Brit. Myc. Soc. **62**: 401–429.

______, and J. Dee. 1967. Plasmodial fusion and lethal interaction between strains in a Myxomycete. Nature (London) **215**: 832–834.

Carter, S., and E. N. Nannenga-Bremekamp. 1972. A new species of *Physarum* (Myxomycetes) with a note on the delimitation of the genera *Physarum* and *Badhamia*. Proc. Ned. Akad. Wet., ser. C, **75**: 326–330.

Charvat, I., I. K. Ross, and J. Cronshaw. 1973. Ultrastructure of the plasmodial slime mold *Perichaena vermicularis*. II. Peridium formation. Protoplasma **78**: 1–20.

______, J. Cronshaw, and I. K. Ross. 1974. Development of the capillitium in *Perichaena vermicularis*, a plasmodial slime mold. Protoplasma **80**: 207–221.

Chassain, M. 1980. Essai sur la place ecologique des Myxomycetes. Docum. Mycol. **11**: 47–57.

Cihlar, C. 1916. Mikrokemijaska istrazivan johitin ublinskim membranama. Bot. Centralbl. **131**: 524.

Clark, J. 1977. Plasmodial incompatibility reactions in the true slime mold *Physarum cinereum*. Mycologia **69**: 46–52.

______, and O. R. Collins. 1973. Directional cytotoxic reactions between incompatible plasmodia of *Didymium iridis*. Genetics **73**: 247–257.

______, and R. Hakim. 1980. Nuclear sieving of *Didymium iridis* plasmodia. Exper. Mycol. **4**: 17–22.

Cohen, A. L. 1939. Nutrition of the Myxomycetes I. Pure culture and two-membered culture of myxomycete plasmodia. Bot. Gaz. **101**: 243–275.

______. 1942. The organization of protoplasm: a possible experimental approach. Growth **6**: 259–272.

______. 1959. An electron microscope study of flagellation in myxomycete swarm cells (Abs.). IX Int. Bot. Congress **2**: 77.

Collins, O. R. 1961. Heterothallism and homothallism in the Myxomycetes. Amer. J. Bot. **48**: 674–683.

______. 1963. Multiple alleles at the incompatibility locus in the Myxomycete *Didymium iridis*. Amer. J. Bot. **50**: 477–480.

______. 1973. Myxomycetes. In Gray, P. (Ed.) Encyclopedia of microscopy and microtechnique, pp. 347–349. Van Nostrand Reinhold Co., New York.

______. 1976. Heterothallism and homothallism: A study of 27 isolates of *Didymium iridis*, a true slime mold. Amer. J. Bot. **63**: 138–143.

______. 1979. Myxomycete biosystematics: Some recent developments and future research opportunities. Bot. Rev. **45**(2): 145–201.

______, and D. A. Betterley. 1982. *Didymium iridis* in past and future research. Chapter 2 (pp. 25–57) in Aldrich, H. C., and J. S. Daniel (l.c., vol. 1).

______, and H.-C. Tang. 1973. *Physarum polycephalum*: pH and plasmodium formation. Mycologia **65**: 232–236.

Considine, J. M., and M. F. Mallette. 1965. Production and partial purification of antibiotic materials formed by *Physarum gyrosum*. Appl. Microbiol. **13**: 464–468.

Cook, O. F. 1902. Types and synonyms. Science **15**: 646–656.

Cooke, M. C. 1877. The Myxomycetes of the United States. Ann. Lyc. Nat. Hist. New York **11**: 378–409.

Copeland, H. F. 1956. The classification of lower organisms. Pacific Books, Palo Alto.

Currie, M. E. 1919. A critical study of the slime-moulds of Ontario. Trans. R. Canad. Inst. **12**: 247–308.

Curtis, D. H. 1968. *Barbeyella minutissima*, a new record for the western hemisphere. Mycologia **60**: 708–710.

Czeczuga, B. 1980. Investigations on carotenoids in fungi VII. Representatives of the Myxomycetes Genus [sic]. Nova Hedw. **32**: 347–352.

Daniel, J. W., J. Kelley, and H. P. Rusch. 1962. Hematin-requiring plasmodial myxomycete. J. Bact. **84**: 1104–1110.

______, and H. P. Rusch. 1962a. Method for inducing sporulation of pure cultures of the Myxomycete *Physarum polycephalum*. J. Bact. **83**: 234–240.

______, and ______. 1962b. Niacin requirement for sporulation of *Physarum polycephalum*. J. Bact. **83**: 1244–1250.

Davis, E. E. 1965. Preservation of Myxomycetes. Mycologia **57**: 986–988.

Dee, J. 1960. A mating type system in an acellular slime mould. Nature (London) **185**: 780–781.

______. 1966. Multiple alleles and other factors affecting plasmodium formation in the true slime mold *Physarum polycephalum*. J. Protozool. **13**: 610–616.

______, in Ashworth, J. M., and J. Dee. 1975. The biology of slime moulds. Stud. in Biol. **#56**. Inst. of Biology, London, England.

Dennison, M. L. 1945. The genus *Lamproderma* and its relationships. I. Mycologia **37**: 80–108.

Diehl, W. W. 1929. An improved method for sealing microscopic mounts. Science **69**: 276–277.

Dhillon, S. S. [1977] 1978. Myxomycetes new to India. I. Sydowia **30**: 1–5.

Dillenius, J. J. 1719. Catal. Plant. c. Gissam nasc. Cum append. Frankfort-am-Main.

Dove, W. F., and H. P. Rusch (eds.). 1980. Growth and Differentiation in *Physarum polycephalum*. 250 pp., Princeton Univ. Press.

Eliasson, U. 1977a. Ecological notes on *Amaurochaete* Rost. (Myxomycetes). Bot. Not. **129**: 419–425.

______. 1977b. Recent advances in the taxonomy of Myxomycetes. Bot. Not. **130**: 483–492.

______. 1981. Ultrastructure of peridium and spores in *Lycogala* and *Reticularia*. Trans. Brit. Mycol. Soc. **77**: 243–249.

______, and N. Lundquist. 1979. Fimicolous Myxomycetes. Bot. Not. **132**: 551–568.

______, and S. Sunhede. 1980. External structure of peridium, pseudocapillitium, and spores in the myxomycete genus *Lycogala* Adans. Bot. Not. **133**: 351–361.

Elliott, E. W. 1949. The swarm cells of Myxomycetes. Mycologia **41**: 141–170.

Ellis, T. T., R. W. Scheetz, and C. J. Alexopoulos. 1973. Ultrastructural observations on capillitial types in the Trichiales (Myxomycetes). Trans. Amer. Microscop. Soc. **92**: 65–79.

Emoto, Y. 1933. Über die in Japan noch nicht bekannten Myxomyceten. III. Bot. Mag. Tokyo **47**: 602–606.

______. 1977. The Myxomycetes of Japan. Sangyo Tosho Publ. Co., Ltd., Tokyo.

Erbisch, F. H. 1964. Myxomycete spore longevity. Michigan Botanist **3**: 120–121.

Esser, R., and R. Blaich. 1973. Heterogenic incompatibility in plants and animals. Adv. Genet. **17**: 107–152.

Evenson, A. 1962. A preliminary report of the Myxomycetes of southern Arizona. Mycologia **53**: 137–144.

Farr, M. L. 1957. A checklist of Jamaican slime-moulds (Myxomycetes). Bull. Inst. Jamaica, sci.ser. **7**: 1–67.

______. 1969. Myxomycetes from Dominica. Contr. U.S. Natl. Herb. **37**: 399–440.

______. 1974. Some new myxomycete records for the Neotropics and some taxonomic problems in the Myxomycetes. Proc. Iowa Acad. **81**: 37–40.

______. 1976a. *Reticularia* Baumgarten (Lichens) versus *Reticularia* Bulliard (Myxomycetes). Taxon **25**: 514.

______. 1976b. Myxomycetes. Flora Neotropica Mon. **#16**. The New York Botanical Garden.

______. 1979. Notes of Myxomycetes II. New taxa and records. Nova Hedwigia **31**: 103–118.

______. 1982. Notes on Myxomycetes III. Mycologia 74: 339–343.

______, and H. W. Keller. 1982. Family Elaeomyxaceae (Myxomycetes) validated. Mycologia **74**: 857–858.

Fitzpatrick, H. M. 1930. The lower fungi. McGraw-Hill Book Co., New York.

Fries, E. M. 1829. Syst. Mycol. vol. **3**. Myxogastres (pp. 67–199). Greifenwald.

Furtado, J. S., and L. S. Olive. 1971. Ultrastructural evidence of meiosis in *Ceratiomyxa fruticulosa*. Mycologia **63**: 413–416.

Gäumann, E. A. 1926, 1949, 1964. Die Pilze, eds. 1, 2, 3. Birkhauser Verlag, Basel.

Gilbert, H. C. 1935. Critical events in the life history of *Ceratiomyxa*. Amer. J. Bot. **22**: 52–74.

———, and G. W. Martin. 1933. Myxomycetes found on the bark of living trees. Univ. Iowa Stud. Nat. Hist. **15**(3): 3–8.

Gleditsch, J. G. 1753. Methodus fungorum. . . . Berlin.

Goodman, E. M. 1972. Axenic culture of myxamoebae of the Myxomycete *Physarum polycephalum*. J. Bact. **III**: 242–247.

———, and H. P. Rusch. 1970. Ultrastructural changes during spherule formation in *Physarum polycephalum*. J. Ultrastruct. Res. 30: 172–183.

Goodwin, D. C. 1961. Morphogenesis of the sporangium of *Comatricha*. Amer. J. Bot. **48**: 148–154.

Gorman, J. A., and A. S. Wilkins. 1980. Developmental phases in the life cycle of *Physarum* and related Myxomycetes. Chapter 6 (pp. 157–202) in Dove, W. F., and H. P. Rusch (l.c.).

Gottsberger, G., and N. E. Nannenga-Bremekamp. 1971. A new species of *Didymium* from Brazil. Proc. Ned. Akad. Wet., ser. C, **74**: 264–268.

Gray, W. D. 1938. The effect of light on the fruiting of Myxomycetes. Amer. J. Bot. **25**: 511–522.

———. 1939. The relation of pH and temperature to the fruiting of *Physarum polycephalum*. Amer. J. Bot. **26**: 709–714.

———. 1941. Some effects of the heterochromatic ultra-violet radiation on myxomycete plasmodia. Amer. J. Bot. **28**: 212–216.

———. 1953. Further studies on the fruiting of *Physarum polycephalum*. Mycologia **45**: 817–824.

———, and C. J. Alexopoulos. 1968. The biology of the Myxomycetes. Ronald Press, New York.

Guttes, E., and S. Guttes. 1960. Pinocytosis in the Myxomycete *Physarum polycephalum*. Exper. Cell Res. **20**: 239–241.

———, S. Guttes, and H. P. Rusch. 1961. Morphological observations on growth and differentiation of *Physarum polycephalum* grown in pure culture. Dev. Biol. **3**: 588–614.

Hagelstein, R. 1942. A new genus of the Mycetozoa. Mycologia **34**: 593–594.

———. 1944. The Mycetozoa of North America based upon the specimens in the herbarium of the New York Botanical Garden. Publ. by the author, Mineola, New York.

Harada, Y. 1977. *Badhamia utricularis* occurring on fruit bodies of *Pholiota nameko* in sawdust culture. Bull. Fac. Agric. Hirosaki Univ. **#28**: 32–42.

Harper, R. A., and B. O. Dodge. 1914. The formation of the capillitium in certain Myxomycetes. Ann. Bot. **28**: 1–18.

Haskins, E. F. 1970. Axenic culture of myxamoebae of the Myxomycete *Echinostelium minutum*. Canad. J. Bot. **48**: 663–664.

———. 1971. Sporophore formation in the Myxomycete *Echinostelium minutum* de Bary. Arch. Protistenk. **113**: 123–129.

———. 1978a. A study on the amoebo-flagellate transformation in the slime mold *Echinostelium minutum* de Bary. Protoplasma **94**: 193–206.

———. 1978b. Vergleich der Plasmodien-Typen und der Sporulation bei Myxomyceten. In Haskins, E. F., N. S. Kerr, and Göttingen Inst. Wiss. Film, biol. sect., ser. 11, **#27**, film C 1220; 34 pp.

Henney, H. R., and P. Chu. 1977. Chemical analyses of cell walls from microcysts and microsclerotia of *Physarum flavicomum*; comparison to slime coat from microplasmodia. Exper. Mycol. **1**: 83–91.

———, and T. Lynch. 1969. Growth of *Physarum flavicomum* and *Physarum rigidum* in chemically defined minimal media. J. Bact. **99**: 531–534.

Henney, M. R. 1967. The mating type system of *Physarum flavicomum*. Mycologia **59**: 637–652.

———, and H. R. Henney, Jr., 1968. The mating-type systems of Myxomycetes *Physarum rigidum* and *P. flavicomum*. J. Gen. Microbiol. **53**: 321–332.

Hertel, R. J. G. 1956. Taxonomia de *Comatricha* Preuss em. Rost. Dusenia **7**: 341–350.

Hinssen, H. 1981. An actin-modulating protein from *Physarum polycephalum*. 2. Calcium-dependence and other properties. Europ. J. Cell Biol. **23**: 234–240.

Höhnel, F. v. 1914. Über *Endrodromia vitrea* Berk. Fragm. z. Mykol. XVI Mitth. #**875**, pp. 97–98. Sitzber. Akad. Wiss. Wien **123**: 145–146.

Holmgren, P. K., and W. Keuken. 1974. Index Herbariorum, pt. 1. The herbaria of the world. Ed. 6. Reg. Veg. vol. **92**. Utrecht, Netherlands.

Honey, N. K., R. T. M. Poulter, and P. J. Winter. 1981. Selfing mutants from heterothallic strains of *Physarum polycephalum*. Genet. Res. **37**: 113–122.

Howard, F. L. 1931. The life history of *Physarum polycephalum*. Amer. J. Bot. **18**: 116–133.

Hung, C.-Y., and L. S. Olive. 1972. Ultrastructure of the spore wall in *Echinostelium*. Mycologia **64**: 1160–1163.

Hüttermann, A. 1973a. *Physarum polycephalum* — Object of Research in cell biology. Ber. Deuts. Bot. Ges. **86**: 1–4.

______. 1973b. Biochemical events during spherule formation of *Physarum polycephalum*. Ber. Deuts. Bot. Ges. **86**: 55–76.

Indira, P. U. 1964. Swarmer formation from plasmodia of Myxomycetes. Trans. Brit. Mycol. Soc. **47**: 531–533.

______. 1966. Studies in Myxomycetes. Unpubl. thesis, University of Madras.

______. 1971. The life cycle of *Stemonitis herbatica*. II. Trans. Brit. Mycol. Soc. **56**: 251–259.

______, and R. Kalyanasundaram. 1963. Preliminary investigations in culture of some Myxomycetes. Ber. Schweiz. Bot. Ges. **73**: 381–386.

Ing, B. 1965. Notes on Myxomycetes. Trans. Brit. Mycol. Soc. **48**: 647–651.

______, and N. E. Nannenga-Bremekamp. 1967. Notes on Myxomycetes. XIII. *Symphytocarpus* nov. gen. Stemonitacearum. Proc. Ned. Akad. Wet., ser. C, **70**: 217–231.

Ingold, C. T. 1939. Spore discharge in land plants. Oxford Univ. Press, Oxford.

Jahn, E. 1899. Zur Kenntnis des Schleimpilzes *Comatricha obtusata* Preuss. Festschr. Schwendener, pp. 288–300. Berlin.

______. [1923] 1924. Myxomycetenstudien. XI. Beobachtungen über seltene Arten. Ber. Deuts. Bot. Ges. **41**: 390–396.

______. 1928. Myxomycetenstudien. 12. Das System der Myxomyceten. Ber. Deuts. Bot. Ges. **46**: 8–17.

______. 1931. Die Stielbildung bei den Sporangien der Gattung *Comatricha*. Ber. Deuts. Bot. Ges. **49**: 77–82.

Jahn, T. L., and E. C. Bovee. 1965. Mechanisms of movement in taxonomy of Sarcodina. I. Amer. Midl. Nat. **73**: 30–40.

Jarocki, J. 1931. Mycetozoa from the Czarnohora mountains in the Polish eastern Carpathians. Bull. Acad. Polon., B, **11**: 447–464.

Jump, J. A. 1954. Studies on sclerotization in *Physarum polycephalum*. Amer. J. Bot. **41**: 561–567.

Kamiya, N. 1950. The protoplasmic flow in the myxomycete plasmodium as revealed by a volumetric analysis. Protoplasma **39**: 344–357.

______. 1959. Protoplasmic streaming. Protoplasmatologia **8**(3a). 199 pp.

Karling, J. S. 1968. The Plasmodiophorales. Ed. 2. Hafner Publ. Co., New York.

Keller, H. W. 1980. Corticolous Myxomycetes VIII. *Trabrooksia*, a new genus. Mycologia **72**: 395–403.

______, H. C. Aldrich, and T. E. Brooks. 1973. Corticolous Myxomycetes II. Notes on *Minakatella longifila* with ultrastructural evidence for its transfer to the Trichiaceae. Mycologia **65**: 768–778.

______, and L. L. Anderson. 1978. Some coprophilous species of Myxomycetes. ASB Bull. **25**: 67.

______, and T. E. Brooks. 1976a. Corticolous Myxomycetes IV: *Badhamiopsis*, a new genus for *Badhamia ainoae*. Mycologia **68**: 834–841.

______, and ______. 1976b. Corticolous Myxomycetes V: Observations on the genus *Echinostelium*. Mycologia **68**: 1204–1220.

______, and ______. 1977. Corticolous Myxomycetes VII. Contribution toward a monograph of *Licea*, five new species. Mycologia **69**: 667–684.

______, and F. Candoussau. 1973. Quelques récoltes rares de Myxomycètes en France. Rev. Mycol. **38**: 114–123.

______, and D. M. Smith. 1978. Dissemination of myxomycete spores through the feeding activities (ingestion-defecation) of an acarid mite. Mycologia **70**: 1239–1246.

Kerr, N. S. 1960. Flagella formation by myxamoebae of the true slime mold *Didymium nigripes*. J. Protozool. **7**: 103–108.
———. 1961. A study of plasmodium formation by the true slime mold, *Didymium nigripes*. Exper. Cell Res. **23**: 603–611.
———. 1965. A simple method of lyophilization for the long-term storage of slime molds and small soil amoebae. BioScience **15**: 469.
———. 1967. Plasmodium formation by a minute mutant of the true slime mold, *Didymium nigripes*. Exper. Cell Res. **45**: 646–655.
———, and M. Sussman. 1958. Clonal development of the true slime mould *Didymium nigripes*. J. Gen. Microbiol. **19**: 173–177.
Kerr, S. T. 1968. Ploidy level in the true slime mold *Didymium nigripes*. J. Gen. Microbiol. **43**: 9–15.
Kirouac-Brunet, J., S. Masson, and D. Pallotta. 1981. Multiple allelism at the *MATB* locus in *Physarum polycephalum*. Canad. J. Genet. Cytol. **23**: 9–16.
Koevenig, J. L. [Techn. Dir.] 1961. Slime Molds. I. Life Cycle (film). Bur. Aud.-Vis. Instr., Ext. Div., Univ. Iowa, Iowa City.
———, and R. C. Jackson. 1966. Plasmodial mitoses and polyploidy in the Myxomycete *Physarum polycephalum*. Mycologia **58**: 662–667.
———, and E. H. Liu. 1981. Carboxymethyl cellulase activity in the myxomycete *Physarum polycephalum*. Mycologia **73**: 1085–1091.
Komnick, H., W. Stockem, and K. E. Wohlfarth-Botterman. 1973. Cell motility: Mechanisms in protoplasmic streaming and amoeboid movement. Internatl. Rev. Cytol. **34**: 169–249.
Kowalski, D. T. [1967] 1968. Observations on the Dianemaceae. Mycologia **59**: 1075–1084.
———. 1969. A new coprophilous species of *Calonema* (Myxomycetes). Madroño **20**: 229–231.
———. 1970. The species of *Lamproderma*. Mycologia **62**: 621–672.
———. 1971. The genus *Lepidoderma*. Mycologia **63**: 490–516.
———. 1972. *Squamuloderma*: A new genus of Myxomycetes. Mycologia **64**: 1282–1289.
———. 1975a. The myxomycete taxa described by Charles Meylan. Mycologia **67**: 448–494.
———. 1975b. The genus *Diacheopsis*. Mycologia **67**: 616–628.
———, and A. A. Hinchee. 1972. *Barbeyella minutissima*: a common alpine Myxomycete, Syesis **5**: 95–97.
Krzemieniewska, H. 1957. A list of Myxomycetes collected in the years 1955–56. Acta Soc. Bot. Polon. **26**: 785–811 (Polish w. Engl. summ.)
———. 1960. Śluzowce Polski. Polska Akad. Nauk Inst. Bot. Warszawa.
Laane, M. M., F. B. Haugli, and T. R. Mellem. 1976. Nuclear behavior during sporulation and germination in the Colonia strain of *Physarum polycephalum*. Norw. J. Bot. **23**: 177–189.
Lakhanpal, T. N., and K. G. Mukerji. 1977. Taxonomic studies on Indian Myxomycetes. V. The genus *Metatrichia* B. Ing. Proc. Indian Nat. Sci. Acad., B, **42**: 125–129.
———, and ———. 1981. Taxonomy of the Indian Myxomycetes. Bibl. Mycol. **#78**.
Leers, J. D. 1789. Flora herbornensis . . . Berlin.
Lieth, H., and G. F. Meyer. 1957. Über den Bau der Pigmentgranula bei den Myxomyceten. Naturwiss. **44**: 449.
Ling, H., and J. Clark. 1981. Somatic cell incompatibility in *Didymium iridis*: locus identification and function. Amer. J. Bot. **68**: 1191–1199.
Link, J. H. F. 1833. Handbuch zur Erkennung der nutzbarsten und am häufigsten vorkommenden Gewächse. 3. Ordo Fungi, subordo **6**. Myxomycetes, pp. 405–422; 432–433. Berlin.
Lister, A. 1894, 1911, and 1925. A monograph of the Mycetozoa . . . . Eds. 1, 2, and 3, the latter two rev. by G. Lister. Brit. Mus. Nat. Hist., London.
———, and G. Lister. 1904. Notes of Mycetozoa from Japan. J. Bot. **42**: 97–99.
Lister, G. 1913. New Mycetozoa. J. Bot. **51**: 1–4.
———. 1921. New or rare species of Mycetozoa. J. Bot. 59: 89–93.
Locquin, M. 1947. Structure du capillitium d'*Hemitrichia serpula*. Compt. Rend. Acad. Sci. (Paris) **224**: 1442–1443.

______. 1948. Culture des Myxomycètes et production de substances antibiotiques par ces champignons. Compt. Rend. Acad. Sci. (Paris) **227**: 149–150.

______. 1949. Récherches sur les simblospores de Myxomycètes. Bull. Soc. Linn. Lyon **18**: 43–46.

Luyet, B. J. 1940. The case against the cell theory. Science **91**: 252–255.

Macbride, T. H. 1892–1893a. The Myxomycetes of eastern Iowa. Bull. Nat. Hist. Univ. Iowa **2**: 99–162; 1892; 384–389. 1893.

______. 1893b. Nicaraguan Myxomycetes. Bull. Nat. Hist. Univ. Iowa **2**: 277–283.

______. 1899 and 1922. The North American Slime-Moulds . . . . Eds. 1 and 2. The MacMillan Co., New York.

______, and G. W. Martin. 1934. The Myxomycetes . . . . The MacMillan Co., New York.

Madelin, M. F., F. Audus, and D. Knowles. 1975. Attraction of plasmodia of the myxomycete, *Badhamia utricularis*, by extracts of the basidiomycete, *Stereum hirsutum*. J. Gen. Microbiol. **89**: 229–234.

Martin, G. W. 1932. Systematic position of the slime molds and its bearing on the classification of the fungi. Bot. Gaz. **93**: 421–435.

______. 1949. Myxomycetes. North Amer. Flora 1(1). The New York Botanical Garden.

______. 1957. Concerning the "cellularity" and acellularity of the Protozoa. Science **125**: 155.

______. [1960] 1961a. The systematic position of the Myxomycetes. Mycologia **52**: 119–129.

______. 1961b. The genus *Schenella*. Mycologia **53**: 25–30.

______. 1962. Taxonomic notes on Myxomycetes, IV. Brittonia **14**: 180–185.

______. 1966. The genera of Myxomycetes. Univ. Iowa Stud. Nat. Hist. **20**(8): 1–32.

______. 1967. *Lycogala exiguum*. Mycologia **59**: 155–160.

______, and C. J. Alexopoulos. 1969. The Myxomycetes. Univ. Iowa Press, Iowa City.

Massee, G. 1892. A monograph of the Myxogastres. London.

McCormick, J. J., J. C. Blomquist, and H. P. Rusch. 1970. Isolation and characterization of a galactoseamine wall from spores and spherules of *Physarum polycephalum*. J. Bact. **104**: 1119–1125.

McManus, M. A. 1961. Laboratory cultivation of *Clastoderma debaryanum*. Amer. J. Bot. **48**: 884–888.

______. 1962. Some observations on plasmodia of the Trichiales. Mycologia **54**: 78–90.

______, and F. McDade. 1961. Cytological studies of the thread phase of *Ceratiomyxa*. Proc. Iowa Acad. Sci. **68**: 79–85.

Meylan, C. 1932. Les espèces nivales du genre *Lamproderma*. Bull. Soc. Vaud. Sci. Nat. **57**: 359–373.

Micheli, P. A. 1729. Nova plantarum genera Iuxta Tournefortii methodum disposita . . . . Florence.

Mims, C. W. 1969. Capillitial formation in *Arcyria cinerea*. Mycologia **61**: 784–798.

______. 1971. An ultrastructural study of spore germination in the Myxomycete *Arcyria cinerea*. Mycologia **63**: 586–601.

______. 1973. A light and electron microscopic study of sporulation in the Myxomycete *Stemonitis virginiensis*. Protoplasma **77**: 35–54.

______, and M. A. Rogers. 1973. An ultrastructural study of spore germination in the Myxomycete *Stemonitis virginiensis*. Protoplasma **78**: 243–254.

______, and ______. 1975. A light and electron microscopic study of stalk formation in the Myxomycete *Arcyria cinerea*. Mycologia **67**: 638–649.

Mitchel, D. H., S. W. Chapman, and M. L. Farr. 1980. Notes on Colorado fungi IV. *Myxomycetes*. Mycotaxon **10**: 299–349.

Mock, D. L., and D. T. Kowalski. 1976. Laboratory cultivation of *Licea alexopouli*. Mycologia **68**: 370–376.

Mohberg, J. 1982. Ploidy throughout the life cycle in *Physarum polycephalum*. Chapter 7 (pp. 253–272) in Aldrich, H. C., and J. W. Daniel (l. c., vol. 1).

Moore, A. R. 1933. On the cytoplasmic framework of the plasmodium of *Physarum polycephalum*. Tohoku Imp. Univ. Sci. Rep. IV. **8**: 189–192.

Morgan, A. P. 1893–1900. The Myxomycetes of the Miami Valley, Ohio. J. Cincinnati Soc. Nat. Hist. I. **15**: 127–143. 1893; II. **16**: 13–36. 1893; III. **16**: 127–156. 1894; IV. **19**: 1–44. 1896; V. **19**: 147–166. 1900.

Müller, O. F. 1777. Flora Danica **4**(12): 1–6.

Nannenga-Bremekamp, N. E. 1962. Notes on Myxomycetes V. On the identity of the genera *Cribraria* and *Dictydium*. Acta Bot. Neerl. **11**: 21–22.
———. 1965. Notes on Myxomycetes IX. The genus *Licea* in the Netherlands. Acta Bot. Neerl. **14**: 131–147.
———. 1967. Notes on Myxomycetes XII. A revision of the Stemonitales. Proc. Ned. Akad. Wet., ser. C, **70**: 201–216.
———. 1974 and 1979. De Nederlandse Myxomyceten. Eds. 1 and 2, plus appendix to 2nd ed. Ned. Natuurhist. Ver. Netherlands.
———, and M. Härkönen. 1979. *Symphytocarpus fusiformis* (Myxomycetes) a new species from Finland. Proc. Ned. Akad. Wet., ser. C, **82**: 371–374.
Nelson, R. K., and M. Orlowski. 1981. Spore germination and swarm cell morphogenesis in the acellular slime mold *Fuligo septica*. Arch. Microbiol. **130**: 189–194.
———, and R. W. Scheetz. 1975. Swarm cell ultrastructure in *Ceratiomyxa fruticulosa*. Mycologia **67**: 733–740.
———, and ———. 1976. Thread phase of *Ceratiomyxa fruticulosa*. Mycologia **68**: 144–150.
———, and ———, and C. J. Alexopoulos. 1977. Elemental composition of *Metatrichia vesparium* sporangia. Mycotaxon **5**: 365–375.
Nygaard, O. F., S. Guttes, and H. P. Rusch. 1960. Nucleic acid metabolism in a slime mold with synchronous mitosis. Biochim. Biophys. Acta **38**: 298–306.
Olive, L. S. 1964. A new member of the Mycetozoa. Mycologia **56**: 885–896.
———. 1967. The Protostelida — a new order of the Mycetozoa. Mycologia **59**: 1–29.
———. 1975. The Mycetozoans. Academic Press, New York.
———, and C. Stoianovitch. 1966. A simple new mycetozoan genus intermediate between *Cavostelium* and *Protostelium*; a new order of Mycetozoa. J. Protozool. **13**: 164–171.
———, and ———. 1971. A minute new *Echinostelium* with protostelid affinities. Mycologia **63**: 1051–1062.
Persoon, C. H. 1794. Neuer Versuch einer systematischen Eintheilung der Schwämme. Neues Mag. Bot. **1**: 63–128.
Pettersson, B. 1940. Experimentelle Untersuchungen über die euanemochore Verbreitung der Sporepflanzen. Acta Bot. Fenn. **25**: 1–103. In Gregory, P. H. 1961. The microbiology of the atmosphere. Interscience Publ. Inc., New York.
Poulter, R. T. M., and J. Dee. 1968. Segregation of factors controlling fusion between plasmodia of the true slime mould *Physarum polycephalum*. Genet. Res. **12**: 71–79.
Quinlan, R. A., A. Roobol, C. I. Pogson, and K. Gull. 1981. A correlation between *in vivo* and *in vitro* effects of the microtubule inhibitors colchicine, parbendazole, and nocodazole on myxamoebae of *Physarum polycephalum*. J. Gen. Microbiol. **122**: 1–6.
Raciborski, M. 1884. Myxomycetum agri cracoviensis genera, species, et varietates novae. Rozp. Akad. Umiej. **12**: 69–86. (in Polish).
Rakoczy, L. 1973. The Myxomycete *Physarum nudum* as a model organism for photobiological studies. Ber. Deuts. Bot. Ges. **86**: 141–164.
Rammeloo, J. 1973. *Trichia arundinariae* sp. nov. (Myxomycetes, Trichiales) from the National Kahuzi Park (Zaïre). Bull. Jard. Bot. Nat. Belg. **43**: 349–352.
———. 1976. Notes concerning the morphology of some myxomycete plasmodia cultured in vitro. Bull. Soc. Bot. Belg. **109**: 195–207.
———. 1978. Systematische Studie van de Trichiales en de Stemonitales (Myxomycetes) van België. Verh. Acad. Wet. België, Kl.Wet., **40**(146). 166 pp.
Raper, K. B. 1973. Acrasiomycetes. In Ainsworth, G. C., F. K. Sparrow, and A. S. Sussman (eds.). The Fungi vol. IVB. Academic Press, New York.
Ray (Rajus), J. 1660. Cat. Plant. c. Cantabrigiam nasc. . . . Cambridge.
Robbrecht, E. 1974. The genus *Arcyria* Wiggers (Myxomycetes) in Belgium. Bull. Jard. Bot. Nat. Belg. **44**: 303–353.
Ross, I. K. [1957a] 1958a. Syngamy and plasmodium formation in the Myxogastres. Amer. J. Bot. **44**: 843–850.
———. [1957b] 1958b. Capillitial formation in the Stemonitaceae. Mycologia **49**: 809–819.
———. [1960] 1961a. Sporangial development in *Lamproderma arcyrionema*. Mycologia **52**: 621–627.
———. 1961b. Further studies on meiosis in the Myxomycetes. Amer. J. Bot. **48**: 244–248.

______. 1964. Pure cultures of some Myxomycetes. Bull. Torrey Bot. Club **91**: 23–31.
______. 1966. Chromosome numbers in pure and gross cultures of Myxomycetes. Amer. J. Bot. **53**: 712–718.
______. 1967a. Syngamy and plasmodium formation in the Myxomycete *Didymium iridis*. Protoplasma **64**: 104–119.
______. 1967b. Growth and development of the Myxomycete *Perichaena vermicularis*. I. Cultivation and vegetative nuclear divisions. Amer. J. Bot. **54**: 617–625.
______. 1973. The Stemonitomycetidae, a new subclass of Myxomycetes. Mycologia **65**: 477–485.
______. 1979. Biology of the fungi. McGraw Hill Book Co., New York.
______, and R. J. Cummings. 1967. Formation of amoeboid cells from the plasmodium of a myxomycete. Mycologia **59**: 725–732.
Rostafinski, J. [T.] 1873. Versuch eines Systems der Mycetozoen. Inaug.-dissert. Strassburg.
______. 1874–1876. Sluzowce (Mycetozoa) Monografia. Towar. Nauk Scist. **5**(4): 1–215. 1874; **6**(1): 217–432. 1875. Dodatek [Appendix] **8**: 1–43. 1876. As this publication is extremely rare, all references are to "Mon." and "Mon. App."
Rupp (Ruppius), H. B. 1718. Flora jenensis. Bailliar, Frankfurt and Leipzig.
Sauer, H. W., K. L. Babcock, and H. P. Rusch. 1969. Sporulation in *Physarum polycephalum*. A model system for studies on differentiation. Exper. Cell Res. **57**: 319–327.
Schaeffer, J. C. 1772–1774. Fungorum qui in Bavaria et Palatinatu c. Ratisbonam nasc. . . . Ed. 2.
Scheetz, R. W. 1972. The ultrastructure of *Ceratiomyxa fruticulosa*. Mycologis **69**: 38–54.
Schinz, H. 1912–1920. Myxogasteres. In Rabenhorst Krypt.-Flora. Ed. 2, **1**(10).
Schoknecht, J. D. 1975. SEM and X-ray microanalysis of calcareous deposits in myxomycete fructifications. Trans. Amer. Microsc. Soc. **94**: 216–223.
______, and H. W. Keller. 1977. Peridial composition of white fructifications in the Trichiales (*Perichaena* and *Dianema*). Canad. J. Bot. **55**: 1807–1819.
Schroeter, J. 1886. Pilze. In Cohn Kryptogamenflora Schlesien **3**(1): 118–133.
Schünemann, E. 1930. Untersuchungen über die Sexualität der Myxomyceten. Planta **9**: 645–672.
Schuster, F. 1964. Electronmicroscope observations on spore formation in the true slime mold *Didymium nigripes*. J. Protozool. **11**: 207–216.
______. 1965. Ultrastructural morphogenesis of solitary stages of true slime molds. Protistologica **1**: 49–62.
Schweinitz, L. D. de. 1822. Synopsis fungorum Carolinae Sup. . . . D. F. Schwaegrichen (ed.). Schr. Nat. Ges. Leipzig **1**: 20–131.
______. 1832. Synopsis fungorum in America boreali. . . . Trans. Amer. Philos. Soc. II, **4**: 141–316.
Simon, H. L., and H. R. Henney, Jr. 1970. Chemical composition of slime from three species of Myxomycetes. FEBS Letters **7**: 80–82.
Skupienski, F. X. 1926. Contribution à l'étude des Myxomycètes en Pologne. Bull. Soc. Myc. France **42**: 142–169.
Smart, R. F. 1937. Influence of certain external factors on spore germination in the Myxomycetes. Amer. J. Bot. **24**: 145–159.
Sobels, J. C. 1950. Nutrition de quelques Myxomycètes en cultures pures et associées et leurs propiétés antibiotiques. Koch & Knuttel, Gouda.
Sparrow, F. K. 1943, 1960. Aquatic Phycomycetes. Eds. 1 and 2. Univ. Michigan Press, Ann Arbor.
______. [1958] 1959. Interrelationships and phylogeny of the aquatic Phycomycetes. Mycologia **50**: 797–813.
Stafleu, F. A. (Ed.) 1978. International Code of Botanical Nomenclature. Regnum Veg. vol. **97**.
Stevens, R. B. (Ed.) 1974. Mycology Guidebook. 703 pp. Publ. for the Mycol. Soc. of America by Washington Press, Seattle.
Stewart, P. A., and B. Stewart. 1960. Electron microscopical studies of plasma membrane formation in slime molds. Norelco Rept. **7**(Jan-Ap): 44–58.
Stosch, H. A. v. 1935. Untersuchungen über die Entwicklungsgeschichte der Myxomyceten. Sexualität und Apogamie bei Didymiaceen. Planta **23**: 623–656.

———. 1937. Über den Generationswechsel der Myxomyceten. Ber. Deuts. Bot. Ges. **55**: 362–369.

———. 1965. Wachstums- und Entwicklungsphysiologie der Myxomyceten. Handb. Pflanzenphysiol. **15**(1): 641–679.

Strasburger, E. 1884. Zur Entwicklungsgeschichte der Sporangien von *Trichia fallax*. Bot. Zeit. **42**: 305–316; 321–326.

Taniguchi, M., K. Kobayashi, and J. Ohta. 1980. Myosin from the myxamoebae of *Physarum polycephalum*. Cell Struct. Funct. **5**: 379–382.

———, K. Yamuzaki, and J. Ohta. 1978. Extraction of contractile protein from myxamoebae of *Physarum polycephalum*. Cell Struct. Funct. **3**: 181–190.

Taylor, R. L., and M. F. Mallette. 1978. Purification of antibiotics from *Physarum gyrosum* by high-pressure liquid chromatography. Prepar. Biochem. **8**: 241–257.

Therrien, C. D. 1966. Microspectrophotometric measurement of nuclear deoxyribonucleic acid content in two Myxomycetes. Canad. J. Bot. **44**: 1667–1675.

———, W. R. Bell, and O. R. Collins. 1977. Nuclear DNA content of myxamoebae and plasmodia in six non-heterothallic isolates of a myxomycete, *Didymium iridis*. Amer. J. Bot. **64**: 286–291.

———, and J. J. Yemma. 1974. Comparative measurements of nuclear DNA in a heterothallic and a self-fertile isolate of the Myxomycete, *Didymium iridis*. Amer. J. Bot. **61**: 400–404.

Thind, K. S. 1977. The Myxomycetes of India. Indian Council Agric. Res. New Delhi.

———, and P. S. Rehill. 1958. The Myxomycetes of the Mussoorie Hills XI. Indian Phytopath. **11**: 96–109.

Tieghem, P. van. 1880. Sur quelques Myxomycètes à plasmode agrégé. Bull. Soc. Bot. France [2] **27**: 317–322.

Torrend, C. 1908. Myxomycètes. Étude des espèces connues jusqu'ici. Brotéria **7**: 5–177. (reprinted 1909).

Ts'o, P. O. P., J. Bonner, L. Eggman, and J. Vinograd. 1956a. Observations on an ATP-sensitive protein system from the plasmodia of a Myxomycete. J. Gen. Physiol. **39**: 325–347.

———, L. Eggman, and J. Vinograd. 1956b. The isolation of myxomyosin, an ATP-sensitive protein from the plasmodium of a myxomycete. J. Gen. Physiol. **39**: 801–812.

———, ———, and ———. 1957a. Physical and chemical studies of myxomyosin, an ATP-sensitive protein in cytoplasm. Biochem. Biophys. Acta **25**: 532–542.

———, ———, and ———. 1957b. The interaction of myxomyosin with ATP. Arch. Biochem. Biophys. **66**: 64–70.

Turnock, G., S. R. Morris, and J. Dee. 1981. A comparison of the proteins of the amoebal and plasmodial phases of the slime mould, *Physarum polycephalum*. Europ. J. Biochem. **115**: 533–538.

Ulrich, R. 1943. Les constituents de la membrane chez les champignons. Rev. Mycol. (Paris). Mem. Hors-Ser. **#3**, 1–44.

Wallroth, C.W.F. 1833. Flora cryptogamia Germaniae 2. Norimbergae. In Compend. Florae German., sect. II, Plantae Cryptog. s. cellulosae, scrips. Math. Jos. Bluff et Carol. Aug. Fingerhuth, vol. IV.

Ward, J. M. 1958. Shift in oxidases with morphogenesis in the slime mold *Physarum polycephalum*. Science **127**: 596.

Watson, S. W., and K. P. Raper. 1957. *Labyrinthula minuta* sp.nov. J. Gen. Microbiol. **17**: 368–377.

Webster, J. 1970. Introduction to fungi. University Press, Cambridge.

Welden, A. L. 1955. Capillitial development in the myxomycetes *Badhamia gracilis* and *Didymium iridis*. Mycologia **47**: 714–728.

Wheals, A. E. 1970. A homothallic strain of the Myxomycete *Physarum polycephalum*. Genetics **66**: 623–633.

Whitney, K. D. 1980. The myxomycete genus *Echinostelium*. Mycologia **72**: 950–987.

Whittaker, R. H. 1969. New concepts of kingdoms of organisms. Science **163**: 150–160.

Wilson, C., and I. K. Ross. 1955. Meiosis in the Myxomycetes. Amer. J. Bot. **42**: 743–749.

Wilson, M., and E. J. Cadman. 1928. The life-history and cytology of *Reticularia lycoperdon*. Trans. R. Soc. Edinb. **55**: 555–608.

Wolf, F. T. 1959. Chemical nature of the photoreceptor pigment inducing fruiting of plasmodia of *Physarum polycephalum*. Photoper. and Rel. Phen. in Pl. & An., pp. 321–326. AAAS, Washington, D.C.

Wormington, W. M., and R. F. Weaver. 1976. Photoreceptor pigment that induces differentiation in the slime mold *Physarum polycephalum*. Proc. Nat. Acad. Sci. U.S.A **73**: 3896–3899.

Youngman, P. J., D. J. Pallotta, B. Hosler, G. Struhl, and C. E. Holt. 1979. A new mating compatibility locus in *Physarum polycephalum*. Genetics **91**: 683–693.

XV. Index

Page numbers in boldface refer to the formal descriptions; those in roman to incidental mention, including keys. Varieties cited in the text are often omitted.

Note: Page numbers for the *Plates* follow the original pagination of the 1969 edition, for purposes of convenient reference.

The Myxomycetes

Plates I-XLI

Plate I

1 *Ceratiomyxa fruticulosa* (Müll.) Macbr.
  a Stalked sporophore, from large group, × 20
  b Tip of branch, × 100
  c Spore, × 1000

2 *Ceratiomyxa morchella* Welden
  a Fructification, × 15
  b Spore, × 1000

3 *Ceratiomyxa sphaerosperma* Boedijn
  a Fructification, × 15
  b Tip of branch, after Boedijn, × 100
  c Spore, × 1000

4 *Licea biforis* Morgan
  a Sporangium, × 5
  b Cluster of three sporangia, × 50
  c Spore, × 1000

5 *Licea castanea* G. Lister
  a Sporangium, × 5
  b Cluster of sporangia, × 20
  c Open sporangium, × 40
  d Sporangial lobes, × 100
  e/f Spores, × 1000

6 *Licea fimicola* Dearn. & Bisby
  a Cluster of sporangia, × 5
  b Cluster of sporangia, × 50
  c Spores, × 1000

7 *Licea kleistobolus* Martin
  a Sporangium, × 5
  b Sporangium, × 100
  c Diagram of lid in section, × 300
  d Spore, × 1000

8 *Licea minima* Fries
  a Sporangium, × 5
  b Two sporangia, × 40
  c Spores, × 1000

9 *Licea operculata* (Wingate) Martin
  a Sporangium, × 10
  b Two sporangia, × 25
  c Spore, × 1000

10 *Licea parasitica* (Zukal) Martin
  a Sporangium, × 5
  b Sporangium, × 50
  c Spore, × 1000

11 *Licea pedicellata* (H. C. Gilbert) H. C. Gilbert
  a Sporangium, × 5
  b Two sporangia, × 50
  c Spore, × 1000

12 *Licea pusilla* Schrad.
  a Sporangium, × 5
  b Two sporangia, × 25
  c Spore in optical section, × 1000
  d Spore from different fruiting, surface view, × 1000

13 *Licea tenera* Jahn
  a Sporangium, × 5
  b Sporangium, × 50
  c Spore, × 1000

14 *Licea tuberculata* Martin
  a Sporangium, × 5
  b Sporangium, × 50
  c Portion of wall, by transmitted light, flattened and showing plates, × 250
  d Spore, × 1000

15 *Licea variabilis* Schrad.
  a Sporangia and plasmodiocarp formed by union of sporangia, × 5
  b Spore, × 1000

Plate I

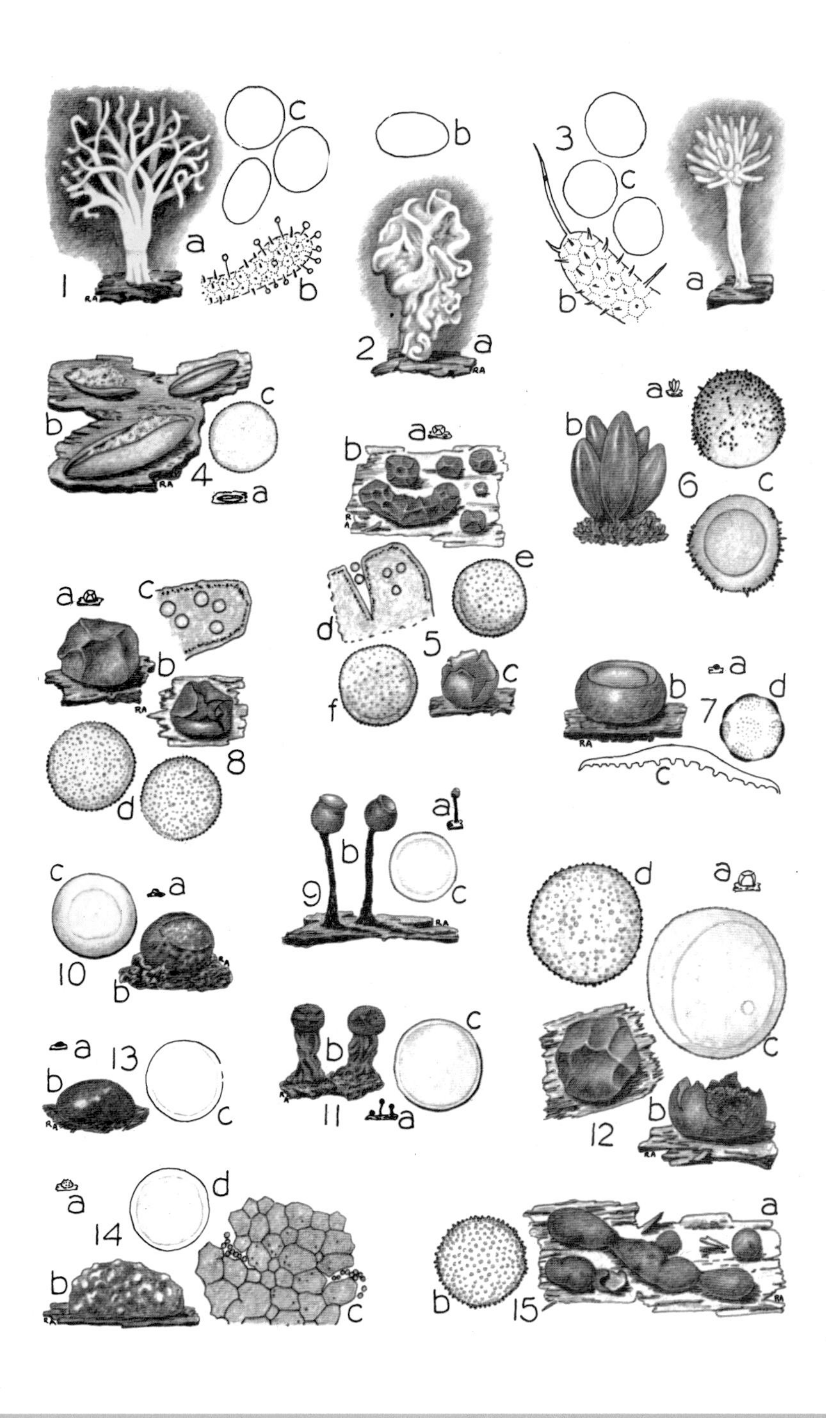

Plate II

16 *Tubifera bombarda* (Berk. & Br.) Martin
- **a** Three sporangia on common stalk, × 10
- **b** Detail of capillitium, showing attachment to stalk, × 25
- **c** Enlarged thread of capillitium, × 500
- **d** Spores, × 1000

17 *Tubifera casparyi* (Rost.) Macbr.
- **a** Pseudoaethalium, in section, × 5
- **b** Columella with branches, × 50
- **c** Spores, × 1000

18 *Tubifera ferruginosa* (Batsch) J. F. Gmel.
- **a** Pseudoaethalium, × 1
- **b** Portion in section, × 10
- **c** Spores, × 1000

19 *Tubifera microsperma* (Berk. & Curt.) Martin
- **a** Pseudoaethalium, × 2
- **b** Spores, × 1000

20 *Tubifera papillata* Martin, Thind and Sohi
- **a** Clustered sporangia on stalk, × 4
- **b** Spores, × 1000

21 *Dictydiaethalium plumbeum* (Schum.) Rost.
- **a** Pseudoaethalium, × 1
- **b** Detail, showing caps and threads, × 20
- **c** Detail of cap with attached threads, × 50
- **d** Spore, × 1000

22 *Lycogala conicum* Pers.
- **a** Four aethalia, × 10
- **b** Spores, × 1000

23 *Lycogala epidendrum* (L.) Fries
- **a** Cluster of aethalia, × 1
- **b** Spores, × 1000
- **c** Thread of pseudocapillitium, × 100

24 *Lycogala exiguum* Morgan
- **a** Cluster of aethalia, × 3
- **b** Detail of peridium, × 200
- **c** Spores, × 1000

25 *Lycogala flavofuscum* (Ehrenb.) Rost.
- **a** Pseudoaethalium, × ½
- **b** Detail of pseudocapillitium, × 100
- **c** Spores, × 1000

Plate II

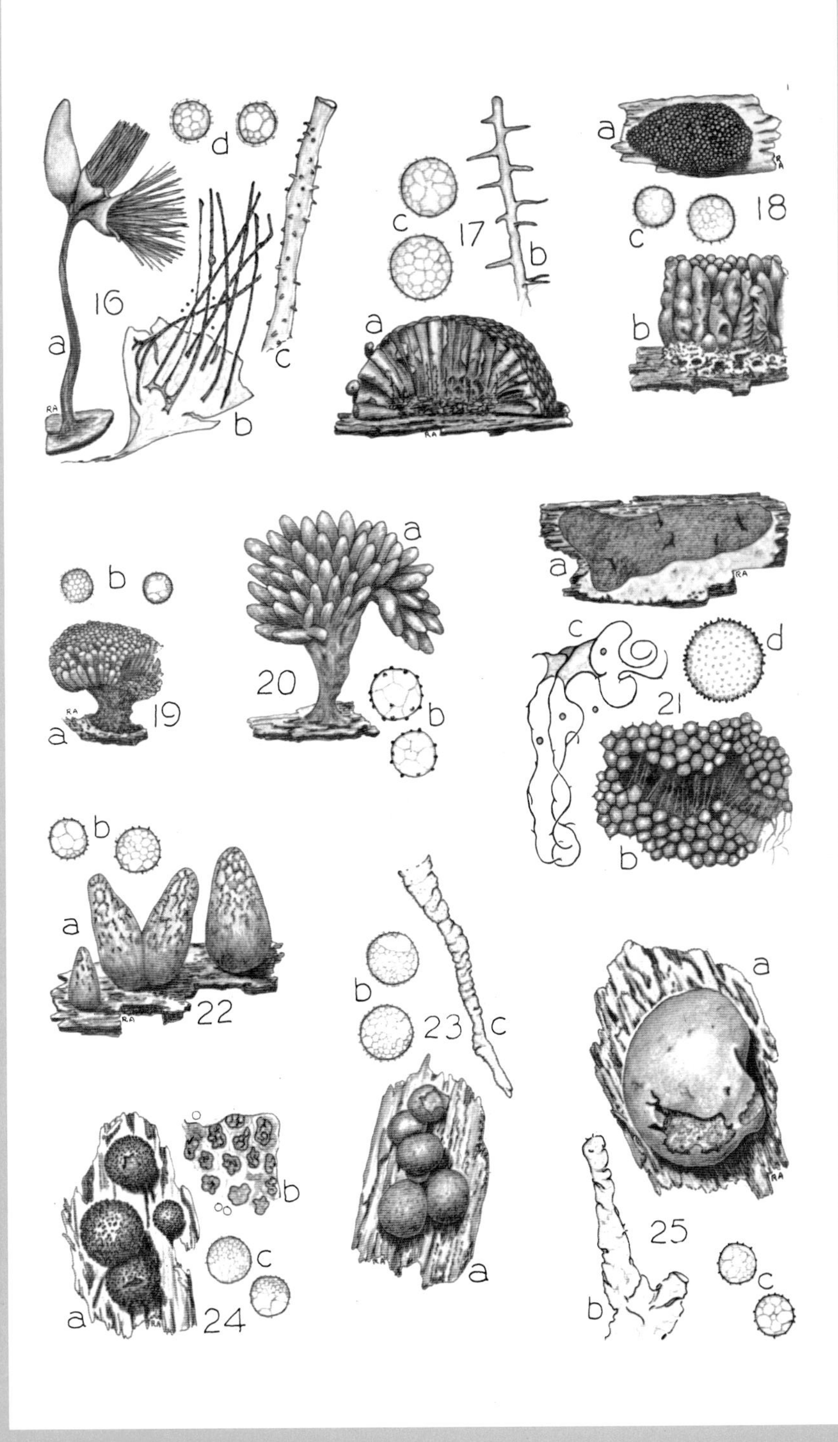

Plate III

26 *Reticularia intermedia* Nann.-Brem.
  a Aethalium, × 1
  b Pseudocapillitium, × 2½
  c Spore, × 1000

27 *Reticularia jurana* Meylan
  a Aethalium, × ½
  b Pseudocapillitium, basal portion at left; outer portion at right, × 50
  c Spore, × 1000

28 *Reticularia lobata* A. Lister
  a Cluster of aethalia, × 2
  b Portion of peridium with attached pseudocapillitium, × 50
  c Spore, × 1000

29 *Reticularia lycoperdon* Bull.
  a Aethalium, × ½
  b Pseudocapillitium, × 3
  c Spores, × 1000

30 *Reticularia olivacea* (Ehrenb.) Fries
  a Aethalium, × 2
  b Pseudocapillitium, × 75
  c Cluster of spores, × 500
  d Isolated spore, × 1000

31 *Reticularia splendens* Morgan
  a Aethalium, × 1
  b Pseudocapillitium, × 25
  c Spore, × 1000

32 *Lindbladia tubulina* Fries
  a Pseudoaethalium, × ½
  b Section through pseudoaethalium with sporangium superimposed, × 5
  c Group of massed sporangia, × 5
  d Two isolated, stipitate sporangia, × 5
  e Spore, with dictydine granules, × 1000

33 *Cribraria argillacea* (Pers.) Pers.
  a Sporangium, × 3
  b Group of sporangia, × 5
  c Sporangium, × 15
  d Detail of surface net, with spores, × 100
  e Spores, with dictydine granules, × 1000

34 *Cribraria atrofusca* Martin & Lovejoy
  a Sporangium, × 3
  b Same, × 15
  c Detail of cup and net, × 50
  d Spore and dictydine granules, × 1000

35 *Cribraria dictyospora* Martin & Lovejoy
  a Sporangium, × 3
  b Same, × 15
  c Spore and dictydine granule, × 1000
  d Portion of surface net, with spores, × 100

36 *Cribraria elegans* Berk. & Curt.
  a Sporangium, × 3
  b Same, × 15
  c Detail of net, with spores, × 100
  d Spore and dictydine granule, × 1000

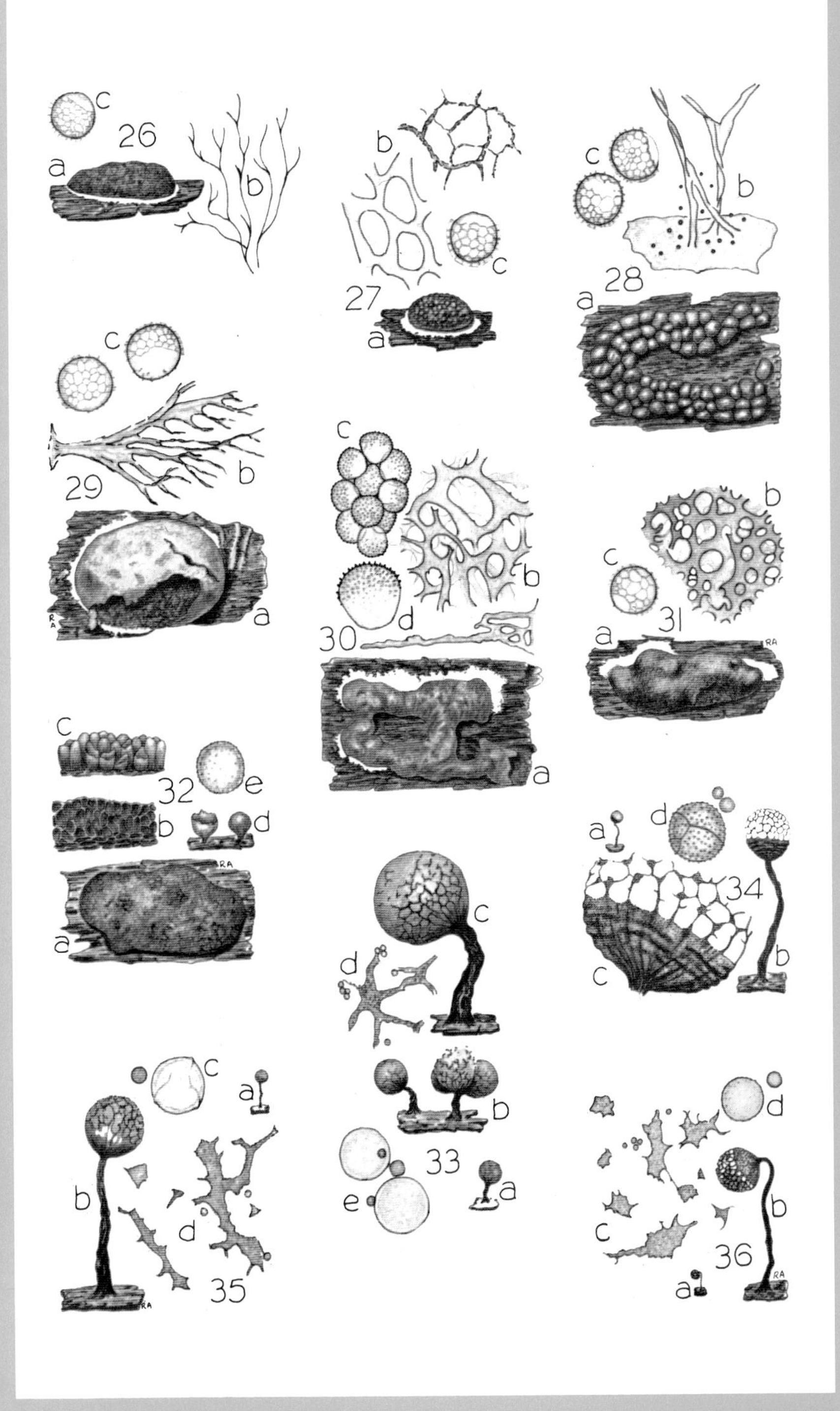
26
27
28
29
30
31
32
33
34
35
36

Plate IV

37 *Cribraria ferruginea* Meylan
- **a** Sporangium, × 3
- **b** Same, × 15
- **c** Detail of cup and net, with spores, × 100
- **d** Spore and dictydine granules, × 1000

38 *Cribraria intricata* Schrad.
- **a** Sporangium, × 3
- **b** Same, with rudimentary cup, × 15
- **c** Same, with developed cup, × 15
- **d** Detail of cup and net, with spores, × 100
- **e** Node, × 250
- **f** Spore and dictydine granules, × 1000

39 *Cribraria languescens* Rex
- **a** Sporangium, × 3
- **b/c** Three sporangia, × 15
- **d** Margin of cup and net, with spores, × 100
- **e** Node, with spore, × 250
- **f** Spore and dictydine granules, × 1000

40 *Cribraria laxa* Hagelst.
- **a** Sporangium, × 3
- **b** Same, × 15
- **c** Net, with spores, × 100
- **d** Node, with spore, × 250
- **e** Spore and dictydine granules, × 1000

41 *Cribraria lepida* Meylan
- **a** Two sporangia, × 3
- **b/c** Two sporangia, × 10
- **d** Node, with spores, × 250
- **e** Spore and dictydine granules, × 1000

42 *Cribraria macrocarpa* Schrad.
- **a** Sporangium, × 3
- **b** Same, × 15
- **c** Detail of net, with spores, × 100
- **d** Spore and dictydine granules, × 1000

43 *Cribraria microcarpa* (Schrad.) Pers.
- **a** Two sporangia, × 3
- **b** Same, × 15
- **c** Detail of net, with spores, × 100
- **d** Node, with spores, × 250
- **e** Spore and dictydine granules, × 1000

44 *Cribraria minutissima* Schw.
- **a** Sporangium, × 3
- **b** Two sporangia, × 15
- **c** Net, with spores, × 100
- **d** Detail of net, with spores, × 500
- **e** Spore and dictydine granules, × 1000

45 *Cribraria oregana* H. C. Gilbert
- **a** Sporangia, × 3
- **b** Two sporangia, × 15
- **c** Detail showing margin of cup and net, with spores, × 100
- **d** Spores and dictydine granules, × 1000

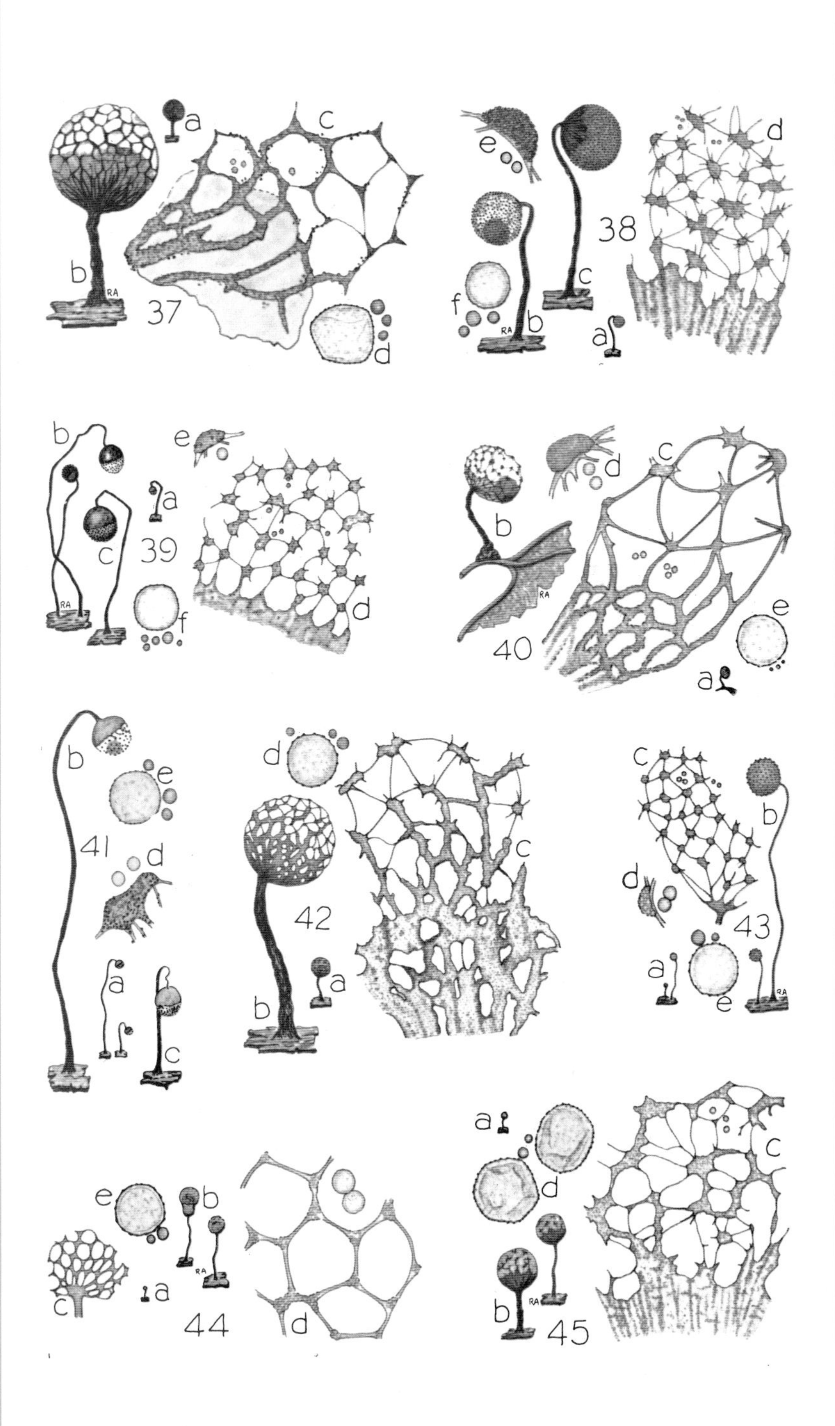
37
a
b
c
d
38
a
b
c
d
e
f
39
a
b
c
d
e
f
40
a
b
c
d
e
41
a
b
c
d
e
42
a
b
c
d
43
a
b
c
d
e
44
a
b
c
d
e
45
a
b
c
d

Plate V

46 *Cribraria piriformis* Schrad.
- **a** Sporangium, × 3
- **b** Two sporangia, the larger containing spores, × 15
- **c** Detail of net, × 100
- **d** Node, × 250
- **e** Spore and dictydine granules, × 1000

47 *Cribraria purpurea* Schrad.
- **a** Sporangium, × 3
- **b** Same, × 15
- **c** Detail of net, × 100
- **d** Spore and dictydine granules, × 1000

48 *Cribraria rubiginosa* Fries
- **a** Sporangium, × 3
- **b** Same, × 10
- **c** Detail of net, × 100
- **d** Spore and dictydine granules, × 1000

49 *Cribraria rufa* (Roth) Rost.
- **a** Two sporangia, × 3
- **b** Same, × 15
- **c** Detail of net, × 100
- **d** Spore and dictydine granule, × 1000

50 *Cribraria splendens* (Schrad.) Pers.
- **a** Two sporangia, × 3
- **b** Same, × 15
- **c** Apex of stipe and net, × 100
- **d** Spore and dictydine granules, × 1000

51 *Cribraria tenella* Schrad.
- **a** Two sporangia, × 3
- **b** Three sporangia, showing variation in stalks and cups, × 10
- **c** Margin of cup with net attached, and portion of net from another sporangium, × 100
- **d** Node, × 250
- **e** Spore and dictydine granules, × 1000

52 *Cribraria violacea* Rex
- **a** Two sporangia, showing variation in size, × 3
- **b** Sporangium, × 25
- **c** Sporangium, × 100
- **d** Spore and dictydine granules, × 1000

53 *Cribraria aurantiaca* Schrad.
- **a** Sporangium, × 3
- **b** Two sporangia, × 15
- **c** Margin of cup and net, × 100
- **d** Node, × 250
- **e** Spore and dictydine granules, × 1000

Plate V

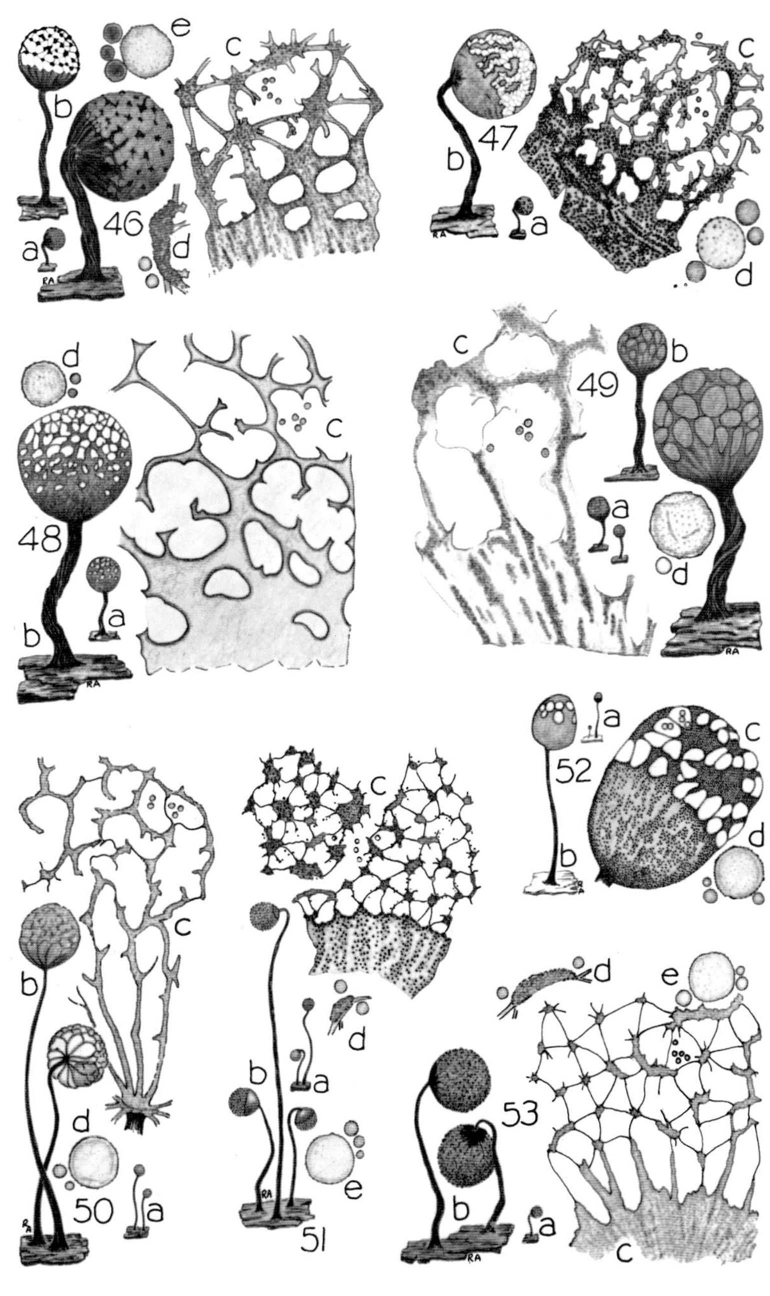

Plate VI

54 *Dictydium cancellatum* (Batsch) Macbr.
a Sporangium, × 20
b Same, without cup, × 40
c Same, with cup, × 40
d Same, with cribrarioid net above, × 40
e Detail of net, × 250
f Spore and dictydine granules, × 1000

55 *Dictydium mirabile* (Rost.) Meylan
a Sporangium, × 20
b Spore and dictydine granule, × 1000

56 *Dictydium rutilum* G. Lister
a Two sporangia, × 20
b Spore and dictydine granules, × 1000

57 *Echinostelium cribrarioides* Alexop.
a Sporangium with net only, × 100
b Spore, × 1000

58 *Echinostelium elachiston* Alexop.
a Sporangium with spores clustered at tip of stalk, × 100
b Spore, × 1000

59 *Echinostelium fragile* Nann.-Brem.
a Stalk with basal disk and columella, × 200
b Tip of same, enlarged, × 500
c Spore, × 1000

60 *Echinostelium minutum* de Bary
a Sporangium with spores clustered at tip, × 100
b Capillitium, × 250
c Spore, × 1000

61 *Listerella paradoxa* Jahn
a Cluster of sporangia, × 10
b Two sporangia, × 50
c Capillitium, peridial lobe and spores, × 500
d Spore, × 1000

62 *Calomyxa metallica* (Berk.) Nieuwl.
a Cluster of fructifications, × 10
b Detail of capillitium, and spores, × 500
c Spore, × 1000

63 *Dianema corticatum* A. Lister
a Sporangium and plasmodiocarp, × 10
b Detail of capillitium, and spores, × 500
c Spore, × 1000

64 *Dianema depressa* (A. Lister) A. Lister
a Sporangia and plasmodiocarp, × 5
b Section through plasmodiocarp, × 5
c Detail of capillitium, and spores, × 500
d Spore, × 1000

65 *Dianema harveyi* Rex
a Sporangia and plasmodiocarp, × 5
b Detail of capillitium, showing attachment to peridium, and spores, × 500
c Spore, × 1000

66 *Prototrichia metallica* (Berk.) Massee
a Sporangium, × 10
b Small stalked sporagium, × 10
c Detail of capillitium, and spores, × 500
d Spore, × 1000

Plate VI

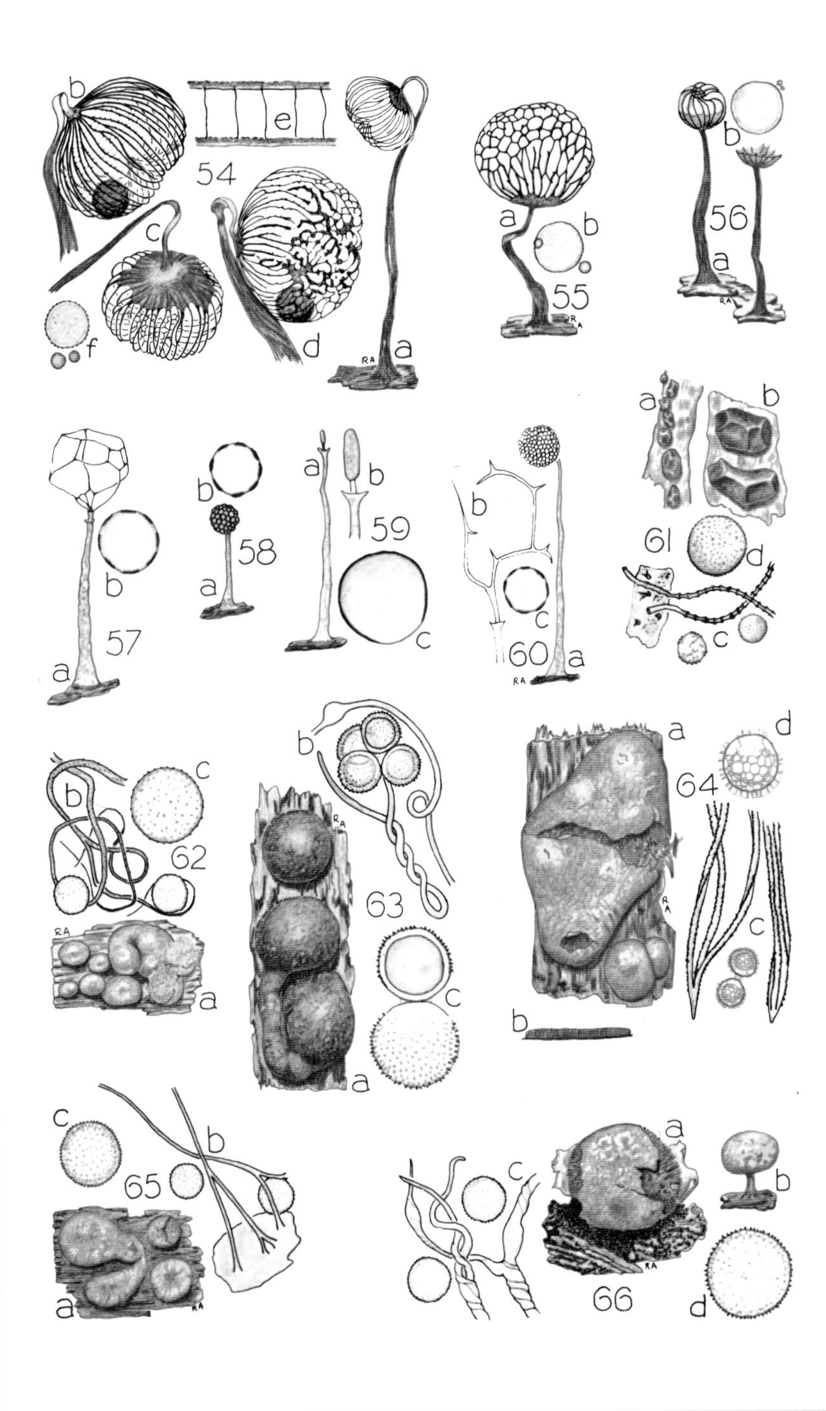

Plate VII

67 *Perichaena chrysosperma* (Currey) A. Lister
- **a** Cluster of fructifications showing variation, × 3
- **b** Semistipitate sporangium, × 10
- **c** Detail of long-spined capillitium, and spore, × 500
- **d** Detail of short-spined capillitium, × 500
- **e** Spore, × 1000

68 *Perichaena corticalis* (Batsch) Rost.
- **a** Cluster of sporangia, × 20
- **b** Detail of capillitium, and spores, × 500
- **c** Spore, × 1000

69 *Perichaena depressa* Libert
- **a** Cluster of sporangia, × 5
- **b** Single sporangium with lid raised by expanding spore-mass, × 20
- **c** Detail of capillitium, and spores, × 500
- **d** Spore, × 1000

70 *Perichaena microspora* Penzig & G. Lister
- **a** Plasmodiocarp, × 5
- **b** Detail of torulose capillitium and spores, × 500
- **c** Detail of spiny capillitium, and spores, × 500
- **d** Spore, × 1000

71 *Perichaena minor* (G. Lister) Hagelst.
- **a** Three sporangia, × 5
- **b** Sporangium, × 30
- **c** Detail of capillitium, and spores, × 500
- **d** Spore, × 1000

72 *Perichaena syncarpon* T. E. Brooks
- **a** Cluster of sporangia, × 10
- **b** Plasmodiocarp, × 20
- **c** Detail of capillitium and cluster of spores, × 500
- **d** Isolated spore, × 1000

73 *Perichaena vermicularis* (Schw.) Rost.
- **a** Plasmodiocarp, × 5
- **b** Detail of capillitium attached to fragment of peridium, and spores, × 500
- **c** Spore, × 1000

74 *Oligonema flavidum* (Peck) Peck
- **a** Cluster of sporangia, × 4
- **b** Same, × 10
- **c** Capillitial threads, and spore, × 500
- **d** Spore, × 1000

75 *Oligonema fulvum* Morgan
- **a** Two sporangia, × 20
- **b** Capillitial thread, and spores, × 500
- **c** Spore in optical section, × 1000

76 *Oligonema schweinitzii* (Berk.) Martin
- **a** Heap of sporangia, × 5
- **b** Portion of same, × 20
- **c** Capillitial thread, and spores, × 500
- **d** Spore, × 1000

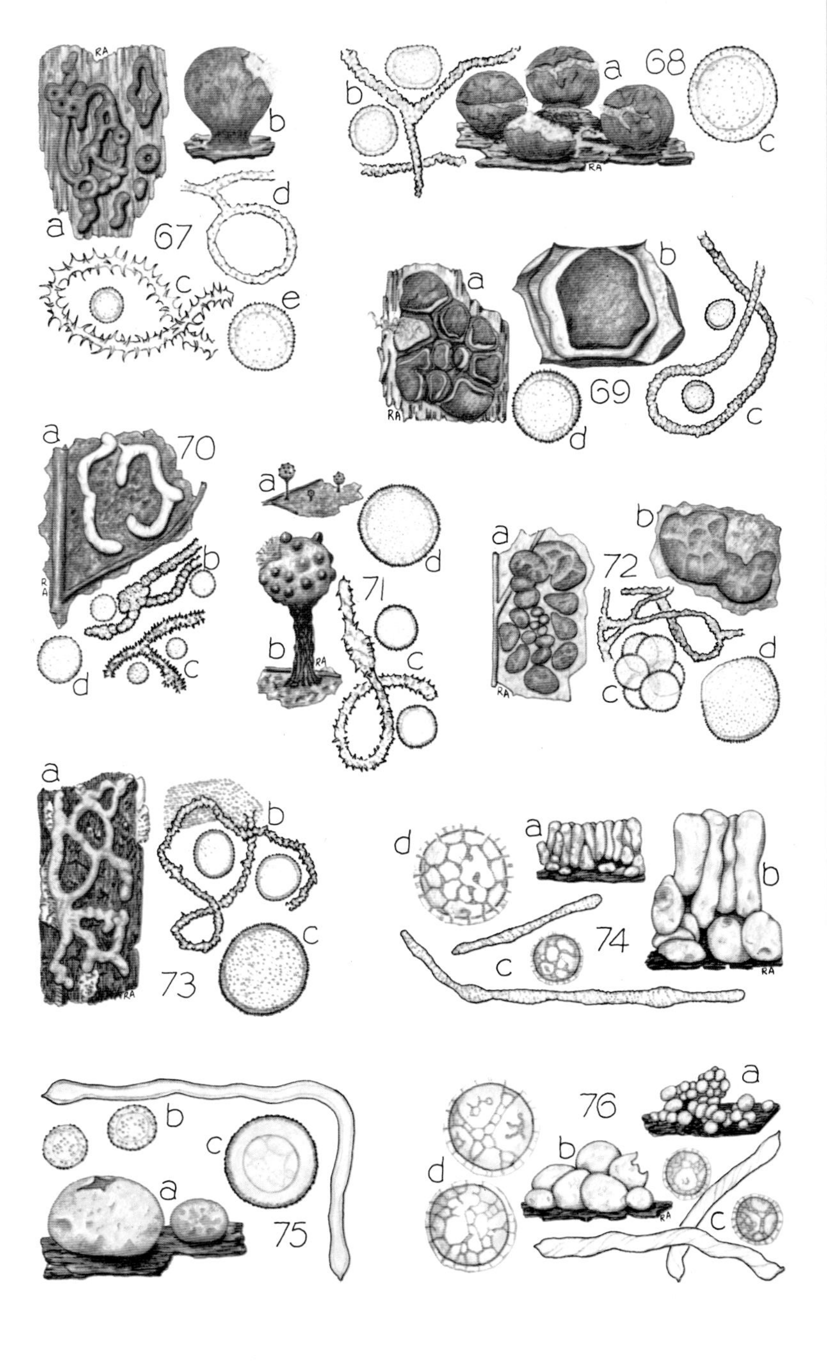

67
a
b
c
d
e
68
a
b
c
69
a
b
c
d
70
a
b
c
d
71
a
b
c
d
72
a
b
c
d
73
a
b
c
74
a
b
c
d
75
a
b
c
76
a
b
c
d

Plate VIII

77 *Calonema aureum* Morgan
- **a** Cluster of sporangia, × 5
- **b** Portion of same, × 20
- **c** Detail of capillitium, and spores, × 500
- **d** Spore, × 1000

78 *Arcyria annulifera* Torrend
- **a** Sporangium, × 10 (in center)
- **b** Same, × 20
- **c** Detail of capillitium, and spores, × 500
- **d** Spore, × 1000

79 *Arcyria carnea* (G. Lister) G. Lister
- **a** Two sporangia and cup of a third, × 10
- **b** Detail of capillitium, and spores, × 500
- **c** Spore, × 1000

80 *Arcyria cinerea* (Bull.) Pers.
- **a** Cluster of digitate gray sporangia, × 10
- **b** Single isolated ochraceous sporangium, × 10
- **c** Four small sporangia, to illustrate variation in size and color, × 10
- **d** Detail of capillitium, and spores, × 500
- **e** Smoother and stouter basal capillitium, × 500
- **f** Spore, × 1000

81 *Arcyria corymbosa* Farr & Martin
- **a** Cluster of sporangia, × 10
- **b** Detail of capillitium, and spores, × 500
- **c** Spore, × 1000

82 *Arcyria denudata* (L.) Wettst.
- **a** Sporangium, × 10
- **b** Detail of capillitium, and spores, × 500
- **c** Spore, × 1000

83 *Arcyria ferruginea* Sauter
- **a** Two sporangia, one with capillitium detached, × 10
- **b** Detail of capillitium, and spores, × 500
- **c** Spore, × 1000

84 *Arcyria glauca* A. Lister
- **a** Sporangium, × 10
- **b** Detail of capillitium, and spores, × 500
- **c** Spore, × 1000

85 *Arcyria globosa* Schw.
- **a** Three sporangia on spine of chestnut bur, × 10
- **b** Detail of capillitium, and spores, × 500
- **c** Spore, × 1000

86 *Arcyria incarnata* (Pers.) Pers.
- **a** Group of sporangia, × 2
- **b** Sporangium, with empty cups, × 10
- **c** Detail of capillitium, and spores, × 500
- **d** Spore, × 1000

Plate VIII

Plate IX

87 *Arcyria insignis* Kalchbr. & Cooke
- **a** Two sporangia, × 10
- **b** Detail of capillitium, and spores, × 500
- **c** Spore, × 1000

88 *Arcyria leiocarpa* (Cooke) Martin & Alexop.
- **a** Two sporangia, × 10
- **b** Detail of capillitium and spores, × 500
- **c** Spore, × 1000

89 *Arcyria magna* Rex
- **a** Cluster of sporangia, × 2
- **b** Base of sporangium, with three empty cups × 10
- **c** Detail of capillitium, and spores, × 500
- **d** Spore, × 1000

90 *Arcyria nutans* (Bull.) Grev.
- **a** Sporangium, × 2
- **b** Same, with two empty cups, × 10
- **c** Detail of capillitium, and spores, × 500
- **d** Spore, × 1000

91 *Arcyria occidentalis* (Macbr.) G. Lister
- **a** Cluster of sporangia, × 10
- **b** Two sporangia, × 20
- **c** Detail of capillitium, and spores, × 500
- **d** Spore, × 1000

92 *Arcyria oerstedtii* Rost.
- **a** Sporangium, × 2
- **b** Base of sporangium, × 10
- **c** Detail of capillitium, and spores, × 500
- **d** Spore, × 1000

93 *Arcyria pomiformis* (Leers) Rost.
- **a** Sporangium, × 10
- **b** Same, × 20
- **c** Detail of capillitium, and spores, × 500
- **d** Spore, × 1000

94 *Arcyria stipata* (Schw.) G. Lister
- **a** Cluster of sporangia, × 5
- **b** Sporangium, × 5
- **c** Isolated sporangium, × 10
- **d** Detail of capillitium, and spores, × 500
- **e** Spore, × 1000

95 *Arcyria versicolor* Phill.
- **a** Two sporangia, at left, capillitium beginning emergence, at right, completely emergent, but still compressed below, × 10
- **b** Detail of capillitium, and spores, × 500
- **c** Spore, × 1000

96 *Arcyria virescens* G. Lister
- **a** Sporangium, with empty cup, × 10
- **b** Detail of capillitium, and spores, × 500
- **c** Spore, × 1000

97 *Arcyodes incarnata* (Alb. & Schw.) O. F. Cook
- **a** Cluster of sporangia, × 5
- **b** Portion of cluster, × 20
- **c** Detail of capillitium showing attachment to peridium, and spores, × 500
- **d** Spore, × 1000

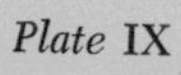
Plate IX

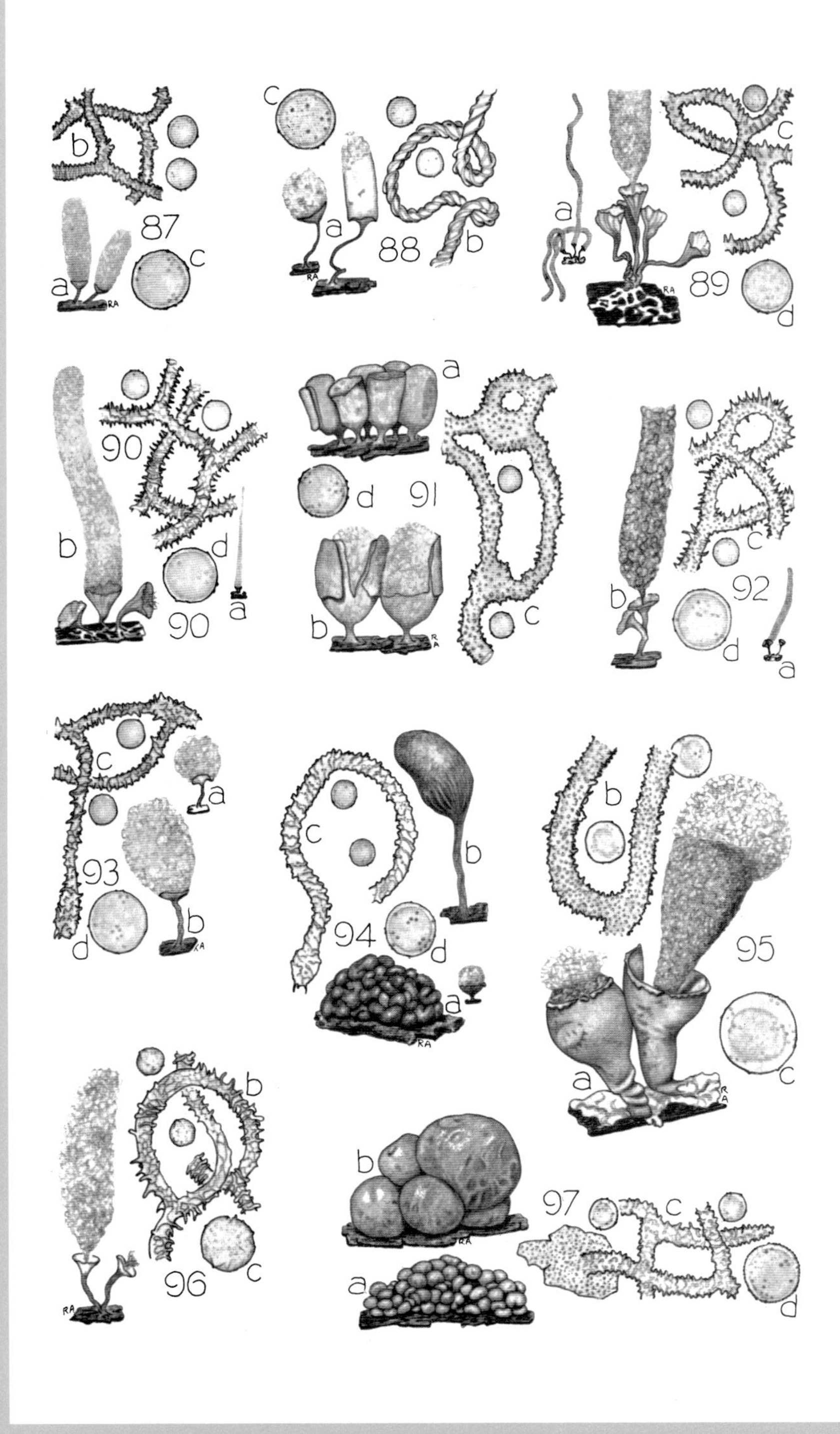
87
a
b
c
88
a
b
c
89
a
c
d
90
a
b
d
91
a
b
c
d
92
a
b
c
d
93
a
b
c
d
94
a
b
c
d
95
a
c
96
b
c
97
a
b
c
d

Plate X

98 *Cornuvia serpula* (Wigand) Rost.
- **a** Cluster of fructifications, × 10
- **b** Plasmodiocarp, × 20
- **c** Detail of capillitium, and spores, × 500
- **d** Spore, × 1000

99 *Trichia alpina* (R. E. Fries) Meylan
- **a** Cluster of fructifications, × 10
- **b** Tips of elaters, and spores, × 500
- **c** Spore, × 1000

100 *Trichia botrytis* (J. F. Gmel.) Pers.
- **a** Cluster of sporangia, × 10
- **b** Elater and spores, × 500
- **c** Spore, × 1000

101 *Trichia contorta* (Ditmar) Rost.
- **a** Cluster of fructifications, × 10
- **b** Tips of elaters and spores, × 500
- **c** Spore, × 1000

102 *Trichia crateriformis* Martin
- **a** Two sporangia, × 5
- **b** Sporangium, × 10
- **c** Elater and spores, × 500
- **d** Spore, × 1000

103 *Trichia erecta* Rex
- **a** Sporangium, × 20
- **b** Tips of elaters and spores, × 500
- **c** Spore, × 1000

104 *Trichia favoginea* (Batsch) Pers.
- **a** Two clusters of sporangia, × 5
- **b** Two sporangia from group, × 20
- **c** Tips of elaters and spores, ×500
- **d** Spore, × 1000

105 *Trichia floriformis* (Schw.) G. Lister
- **a** Sporangia, × 10
- **b** Elater and spores, × 500
- **c** Spore, × 1000

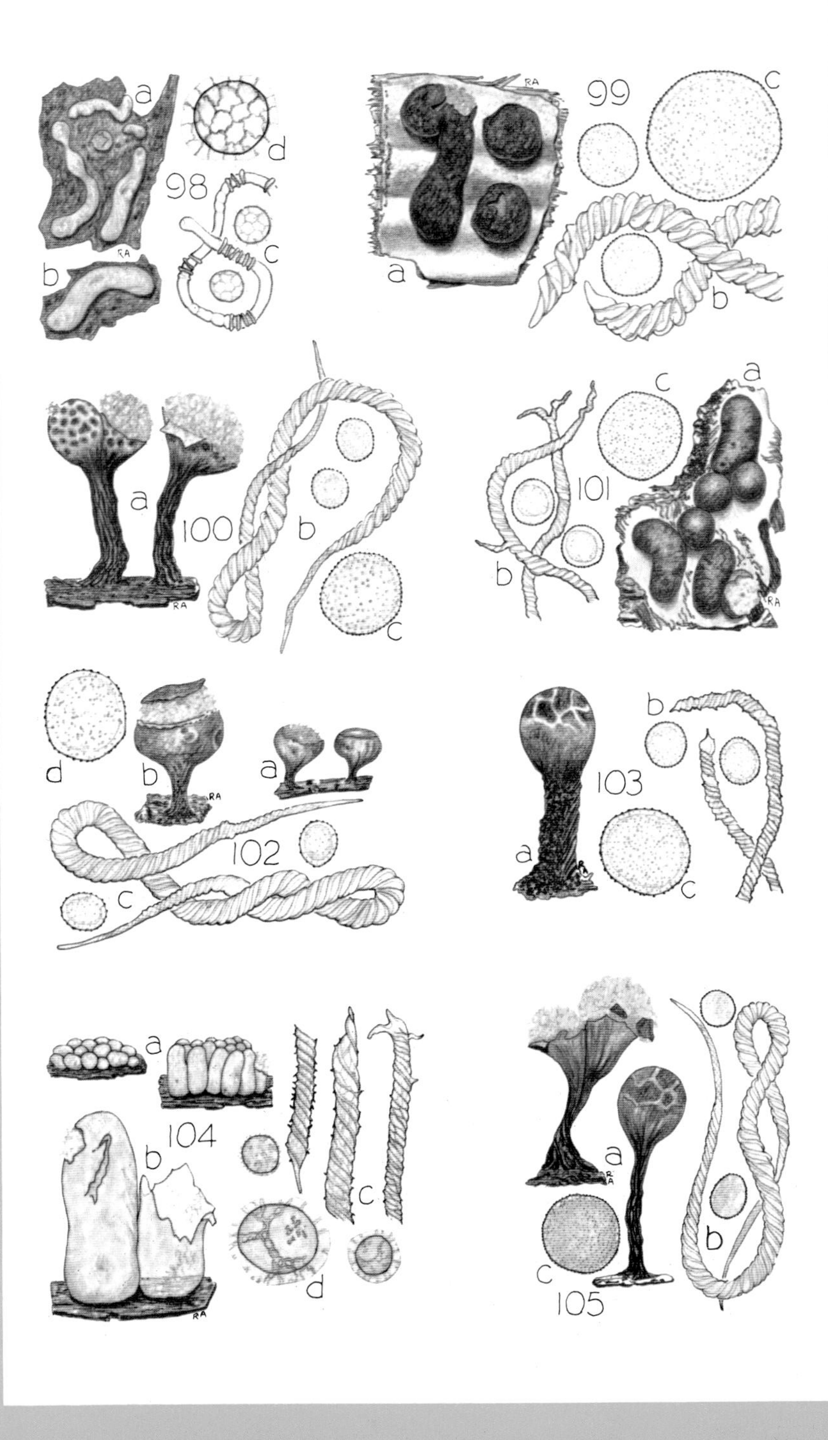

98
a
b
c
d
99
a
b
c
100
a
b
c
101
a
b
c
102
a
b
c
d
103
a
b
c
104
a
b
c
d
105
a
b
c

Plate XI

106 *Trichia lutescens* (A. Lister) A. Lister
- **a** Three sporangia, × 10
- **b** Tips of elaters and spores, × 500
- **c** Spore, × 1000

107 *Trichia macbridei* M. E. Peck
- **a** Cluster of sporangia, × 5
- **b** Pulvinate sporangium, × 10
- **c** Elater and spores, × 500
- **d** Spore, × 1000

108 *Trichia decipiens* (Pers.) Macbride
- **a** Three sporangia, × 10
- **b** Elater and spores, × 500
- **c** Spore, × 1000

109 *Trichia scabra* Rost.
- **a** Cluster of sporangia, × 5
- **b** Three sporangia, × 10
- **c** Elater and spores, × 500
- **d** Spore, × 1000

110 *Trichia subfusca* Rex
- **a** Four sporangia, on three stalks, ×10
- **b** Sporangium, × 20
- **c** Tips of elaters and spores, × 500
- **d** Spore, × 1000

111 *Trichia varia* (Pers.) Pers.
- **a** Cluster of sporangia, × 5
- **b** Stalked sporangium, × 20
- **c** Tips of elaters and spores, × 500
- **d** Spore, × 1000

112 *Trichia verrucosa* Berk.
- **a** Cluster of sporangia on united stalks, × 10
- **b** Tips of elaters and spores, × 500
- **c** Spore, × 1000

113 *Hemitrichia abietina* (Wigand) G. Lister
- **a** Five sporangia, × 10
- **b** Sporangium, × 20
- **c** Free tip of capillitium and spores, × 500
- **d** Spore, × 1000

Plate XI

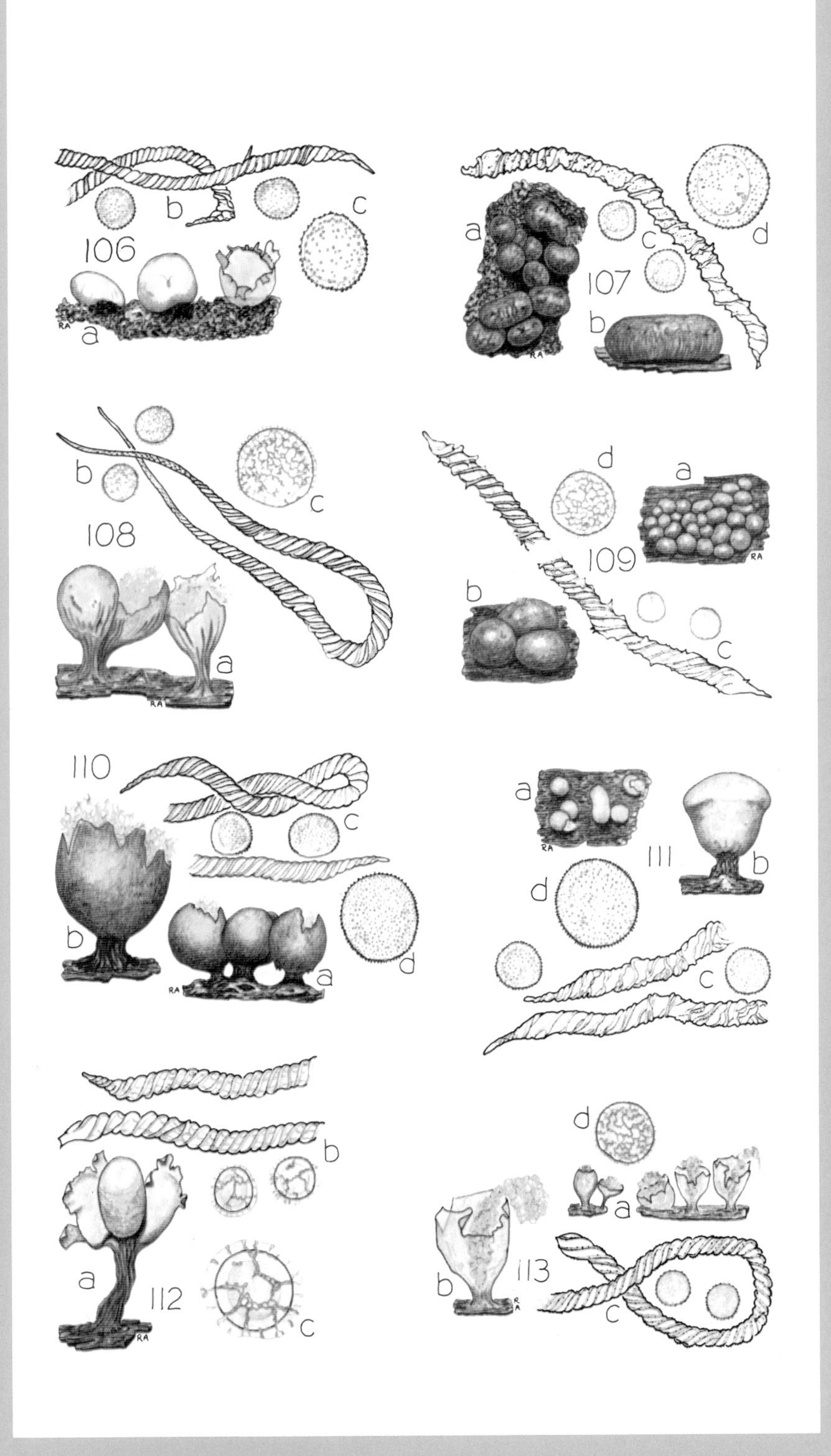

106
a
b
c
107
a
b
c
d
108
a
b
c
109
a
b
c
d
110
a
b
c
d
111
a
b
c
d
112
a
b
c
113
a
b
c
d

Plate XII

114 *Hemitrichia clavata* (Pers.) Rost.
- **a** Three sporangia, × 10
- **b** Detail of capillitium, and spore, × 500
- **c** Spore, × 1000

115 *Hemitrichia intorta* (A. Lister) A. Lister
- **a** Cluster of sessile sporangia, × 10
- **b** Two sporangia, × 25
- **c** Detail of capillitium, and spores, × 500
- **d** Spore, × 1000

116 *Hemitrichia karstenii* (Rost.) A. Lister
- **a** Cluster of fructifications, × 10
- **b** Details of capillitium, and spores, × 500
- **c** Spore, × 1000

117 *Hemitrichia montana* (Morgan) Macbr.
- **a/b** Two sporangia, × 10
- **c** Detail of capillitium including free tip, and spores, × 500
- **d** Spore, × 1000

118 *Hemitrichia paragoga* Farr
- **a** Two sporangia, × 20
- **b** Detail of capillitium, and spores, × 500
- **c** Spore, × 1000

119 *Hemitrichia serpula* (Scop.) Rost.
- **a** Plasmodiocarp, × 10
- **b** Detail of capillitium, and spores, × 500
- **c** Spore, × 1000

120 *Hemitrichia stipitata* (Massee) Macbr.
- **a** Sporangium, × 10
- **b** Detail of capillitium, and spores, × 500
- **c** Spore, × 1000

121 *Metatrichia vesparium* (Batsch) Nann.-Brem.
- **a** Two free sporangia, × 10
- **b** Group of sporangia clustered on united stalks, × 10
- **c** Group of empty sporangia on united stalks, × 10
- **d** Detail of capillitium, and spores, × 500
- **e** Spore, × 1000

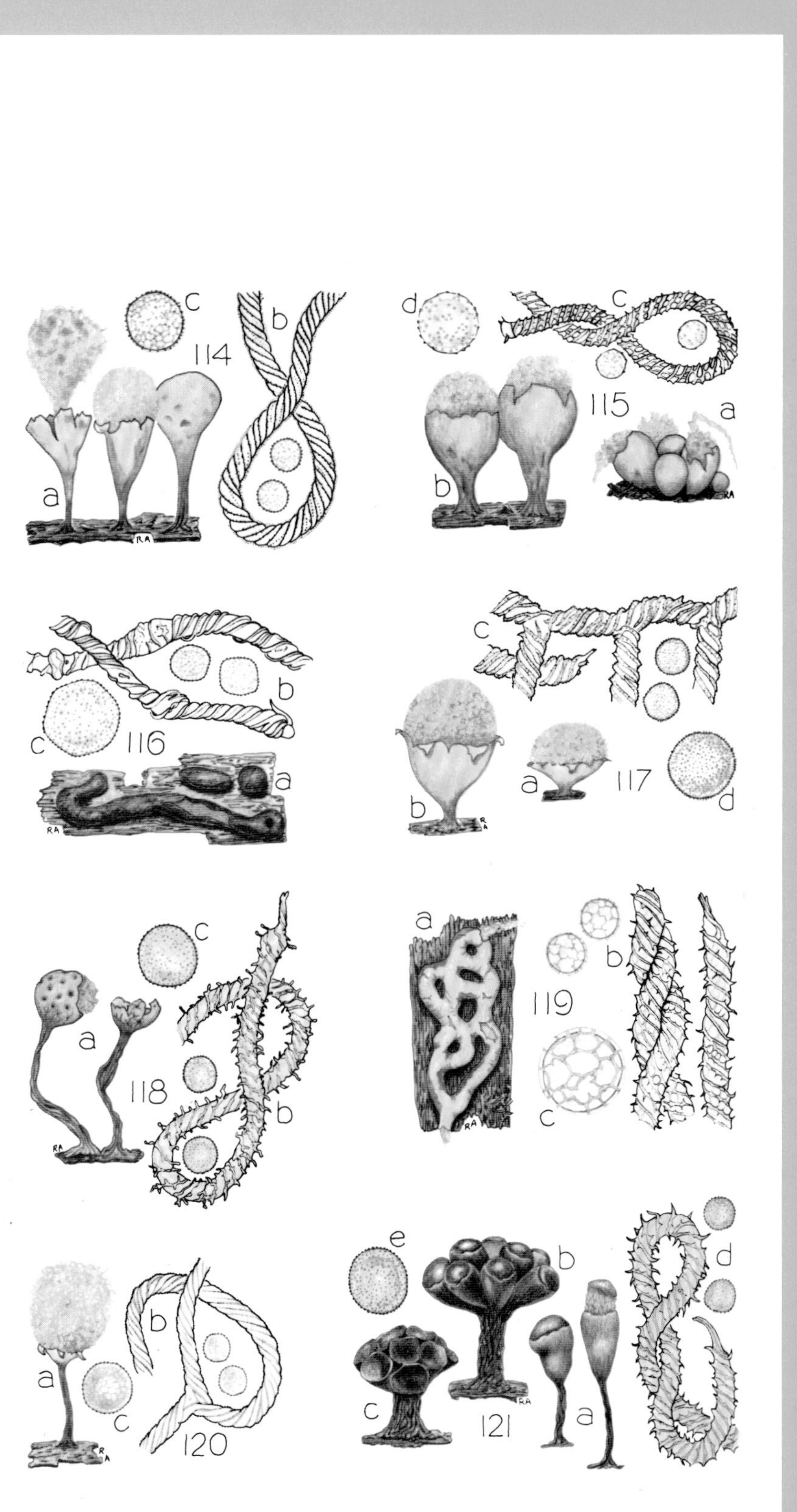
114
a
b
c
115
a
b
c
d
116
a
b
c
117
a
b
c
d
118
a
b
c
119
a
b
c
120
a
b
c
121
a
b
c
d
e

Plate XIII

122 *Colloderma oculatum* (Lippert) G. Lister
- **a** Sporangium, dry, × 20
- **b** Sporangium, wet, showing gelatinous outer wall, × 20
- **c** Small sporangium, × 20, showing inner layer of peridium
- **d** Detail of capillitium, and spores, × 500
- **e** Spore, × 1000

123 *Colloderma robustum* Meylan
- **a** Spore, × 1000

124 *Brefeldia maxima* (Fries) Rost.
- **a** Small aethalium, × 1
- **b** Vesicle and spores, × 500
- **c** Spore, × 1000

125 *Amaurochaete comata* G. Lister & Brândză
- **a** Capillitium and spores, × 100
- **b** Detail of capillitium, and spore, × 500
- **c** Spore, × 1000

126 *Amaurochaete ferruginea* Macbr. & Martin
- **a** Stalk-like branch, from base, × 100
- **b** Tip of same, and spores, × 500
- **c** Spore, × 1000

127 *Amaurochaete atra* (Alb. & Schw.) Rost.
- **a** Two small aethalia, × 1
- **b** Stalk from base giving rise to capillitium, × 10
- **c** Detail of capillitium, and spores, × 500
- **d** Spore, × 1000

128 *Amaurochaete trechispora* Macbr. & Martin
- **a** Capillitium and spores, × 500
- **b** Spore, × 1000

129 *Amaurochaete tubulina* (Alb. & Schw.) Macbr.
- **a** Capillitium and spores, × 500
- **b** Spore, × 1000

130 *Elaeomyxa cerifera* (G. Lister) Hagelst.
- **a** Two sporangia, × 15
- **b** Capillitium and spores, × 500
- **c** Spore, × 1000

131 *Elaeomyxa miyazakiensis* (Emoto) Hagelst.
- **a** Two sporangia, × 15
- **b** Capillitium and spores, × 500
- **c** Spore, × 1000

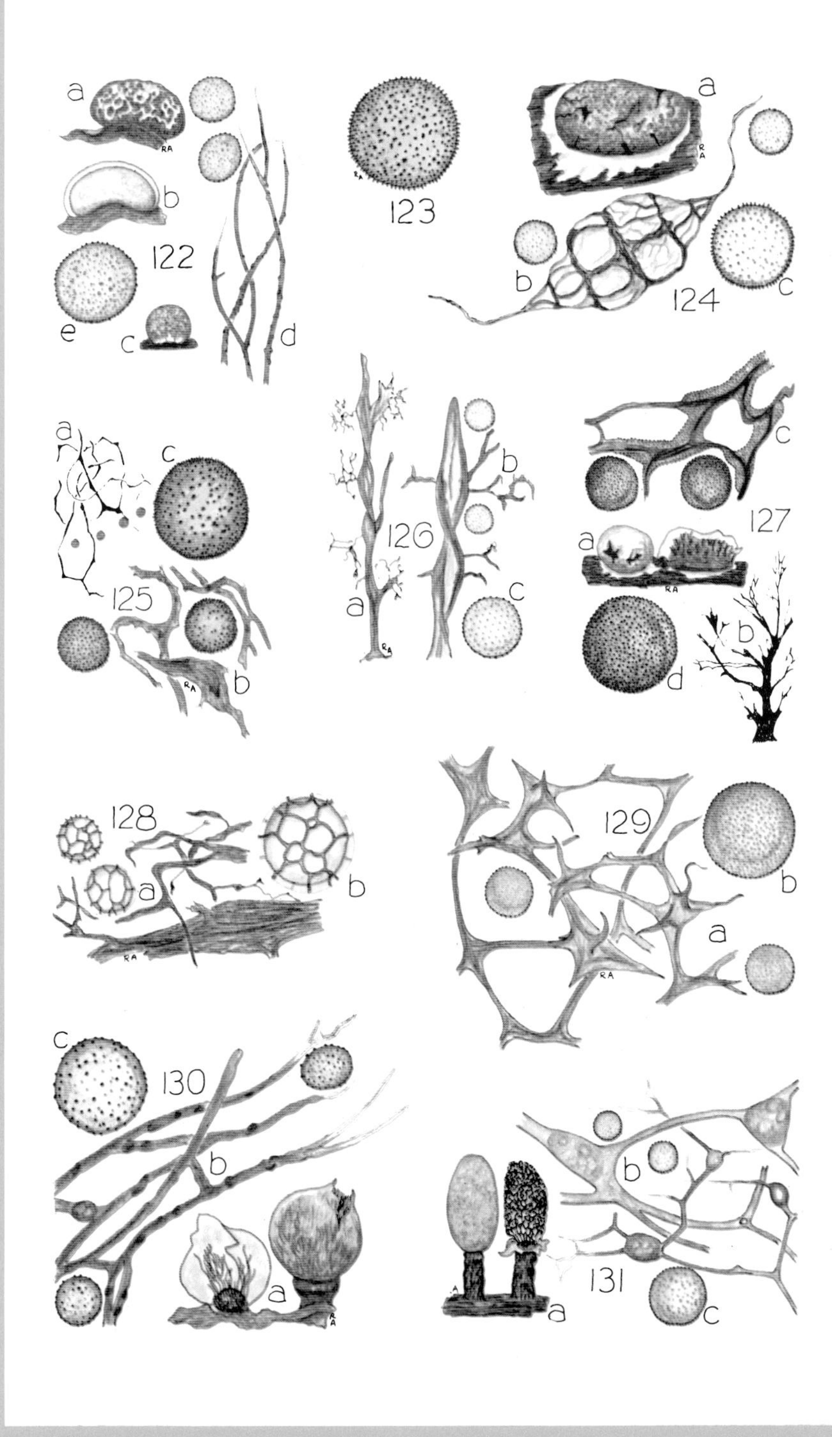
a
b
122
e
c
d
123
a
b
124
c
a
c
125
b
126
a
b
c
c
127
a
b
d
128
a
b
129
b
a
c
130
b
a
b
131
a
c

Plate XIV

132 *Diachea bulbillosa* (Berk. & Br.) A. Lister
- **a** Two sporangia, × 20
- **b** Detail of capillitium, and spores, × 500
- **c** Lime crystals from interior of stalk, × 500
- **d** Spore, × 1000

133 *Diachea leucopodia* (Bull.) Rost.
- **a** Two sporangia, × 20
- **b** Tip of columella surrounded by capillitium, × 50
- **c** Detail of capillitium, and spores, × 500
- **d** Spore, × 1000

134 *Diachea radiata* G. Lister & Petch
- **a** Long-stalked sporangium, × 20
- **b** Short-stalked sporangium, × 20
- **c** Detail of capillitium, with lime crystals from stalk, and spores, × 500
- **d** Spore, × 1000

135 *Diachea splendens* Peck
- **a** Sporangium, × 20
- **b** Detail of capillitium, and spores, × 500
- **c** Spore, × 1000

136 *Diachea subsessilis* Peck
- **a** Long-stalked sporangium, × 20
- **b** Short-stalked and sessile sporangia, × 20
- **c** Detail of capillitium, with spore, and lime crystal from stalk, × 500
- **d** Spore, × 1000

137 *Diachea thomasii* Rex
- **a** Two sporangia, × 20
- **b** Detail of capillitium, and spores, × 500
- **c/d** Spore, × 1000

138 *Schenella microspora* Martin
- **a** Detail of capillitium, and spores, × 500
- **b** Spore, × 1000

139 *Schenella simplex* Macbride
- **a** Pseudoaethalium, × ½
- **b** Diagrammatic longitudinal section, showing sporangia attached to base and cortex, × 3 (a and b after Macbride)
- **c** Cap and base of sporangium, × 20
- **d** Detail of capillitium, and spores, × 500
- **e** Spore, × 1000

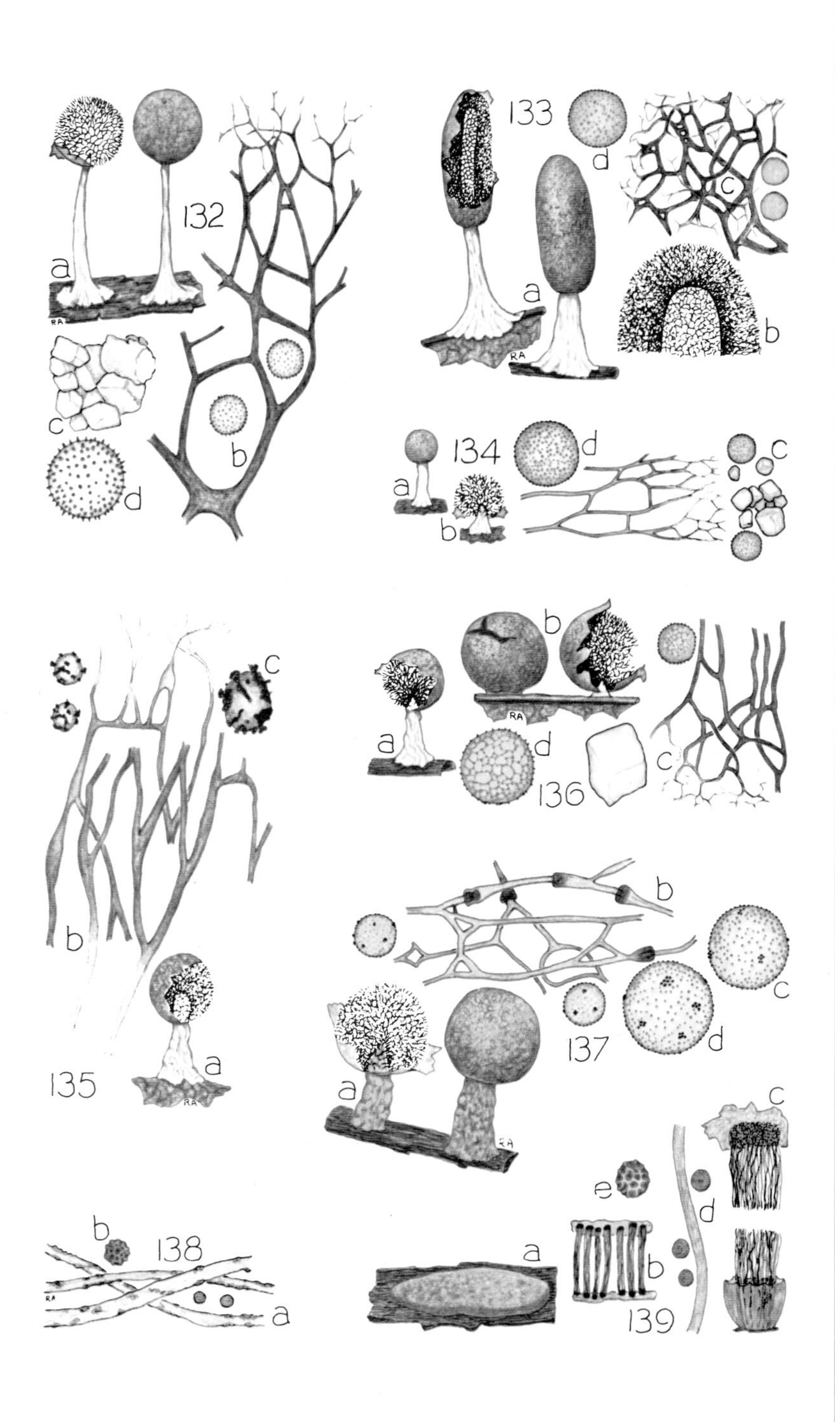
132
a
b
c
d
133
a
b
c
d
134
a
b
c
d
135
a
b
c
136
a
b
c
d
137
a
b
c
d
138
a
b
139
a
b
c
d
e

Plate XV

140 *Enerthenema berkeleyanum* Rost.
- **a** Sporangium, spores shed, × 20
- **b** Tip of same, showing apical cup, × 250
- **c** Two spore clusters, × 500
- **d** Spore cluster, × 1000
- **e** Isolated spore, × 1000

141 *Enerthenema melanospermum* Macbr. & Martin
- **a** Sporangium, spores shed, × 20
- **b** Detail of capillitium, and spores, × 500
- **c** Spore, × 1000

142 *Enerthenema papillatum* (Pers.) Rost.
- **a** Sporangium, × 10
- **b** Same, spores shed, × 20
- **c** Tips of capillitium, and spores, × 500
- **d** Spore, × 1000

143 *Stemonitis axifera* (Bull.) Macbride
- **a** Cluster of sporangia, × 2
- **b** Detail of capillitium, with spores, × 500
- **c** Spore, × 1000

144 *Stemonitis confluens* Cooke & Ellis
- **a** Part of cluster of sporangia, × 2
- **b** Group of united sporangia, × 10
- **c** Detail of capillitium, with spores and membrane, × 250
- **d** Detail of capillitium, with spores, × 500
- **e** Spore, × 1000

145 *Stemonitis flavogenita* Jahn
- **a** Two sporangia, × 2
- **b** Sporangium, × 5
- **c** Tip of sporangium, × 25
- **d** Detail of surface net, and spores, × 500
- **e** Spore, × 1000

146 *Stemonitis fusca* Roth
- **a** Two sporangia, to show range in size, × 2
- **b** Part of a tuft of sporangia, × 2
- **c** Detail of capillitium, with surface net, and spores, × 500
- **d** Spore, × 1000

Plate XV

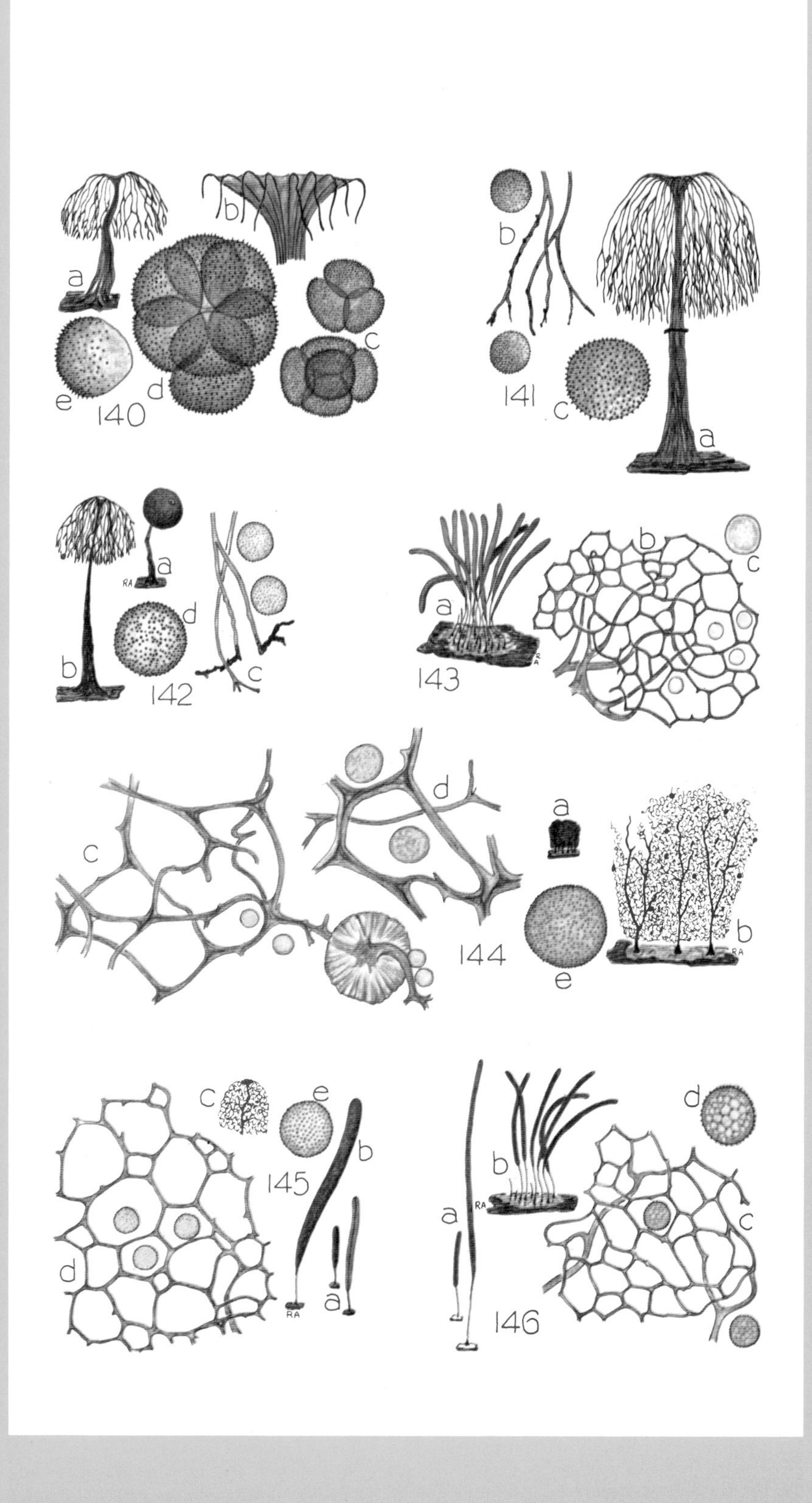

a
b
c
d
e
140
b
141
c
a
a
b
d
142
c
a
b
c
143
c
d
a
144
b
e
c
e
b
145
d
a
d
b
a
c
146

Plate XVI

147 *Stemonitis herbatica* Peck
- **a** Two sporangia, × 2
- **b** Base and upper portion of sporangium, × 25
- **c** Detail of surface net, with spores, × 500
- **d** Spore, × 1000

148 *Stemonitis hyperopta* Meylan
- **a** Four sporangia, × 2
- **b** Sporangium, × 10
- **c** Detail of surface net, and spores, × 500
- **d** Spore, × 1000

149 *Stemonitis mussooriensis* Martin, Thind & Sohi
- **a** Sporangium, × 2
- **b** Same, × 10
- **c** Detail of surface net, with spores, × 500
- **d** Spore, × 1000

150 *Stemonitis nigrescens* Rex
- **a** Three sporangia, × 2
- **b** Sporangium, × 10
- **c** Detail of surface net, with spores, × 500
- **d** Spore, × 1000

151 *Stemonitis pallida* Wingate
- **a** Part of a cluster of sporangia and an isolated sporangium, to show range in size, × 2
- **b** Base and upper portion of sporangium, × 25
- **c** Detail of surface net, and spores, × 500
- **d** Spore, × 1000

152 *Stemonitis smithii* Macbr.
- **a** Sporangia, × 2
- **b** Same, × 10
- **c** Detail of surface net, and spores, × 500
- **d** Spore, × 1000

153 *Stemonitis splendens* Rost.
- **a** Tuft of short sporangia, × 2
- **b** Same, of large sporangia, × 2
- **c** Detail of surface net, with spores, × 500
- **d** Spore, × 1000

Plate XVI

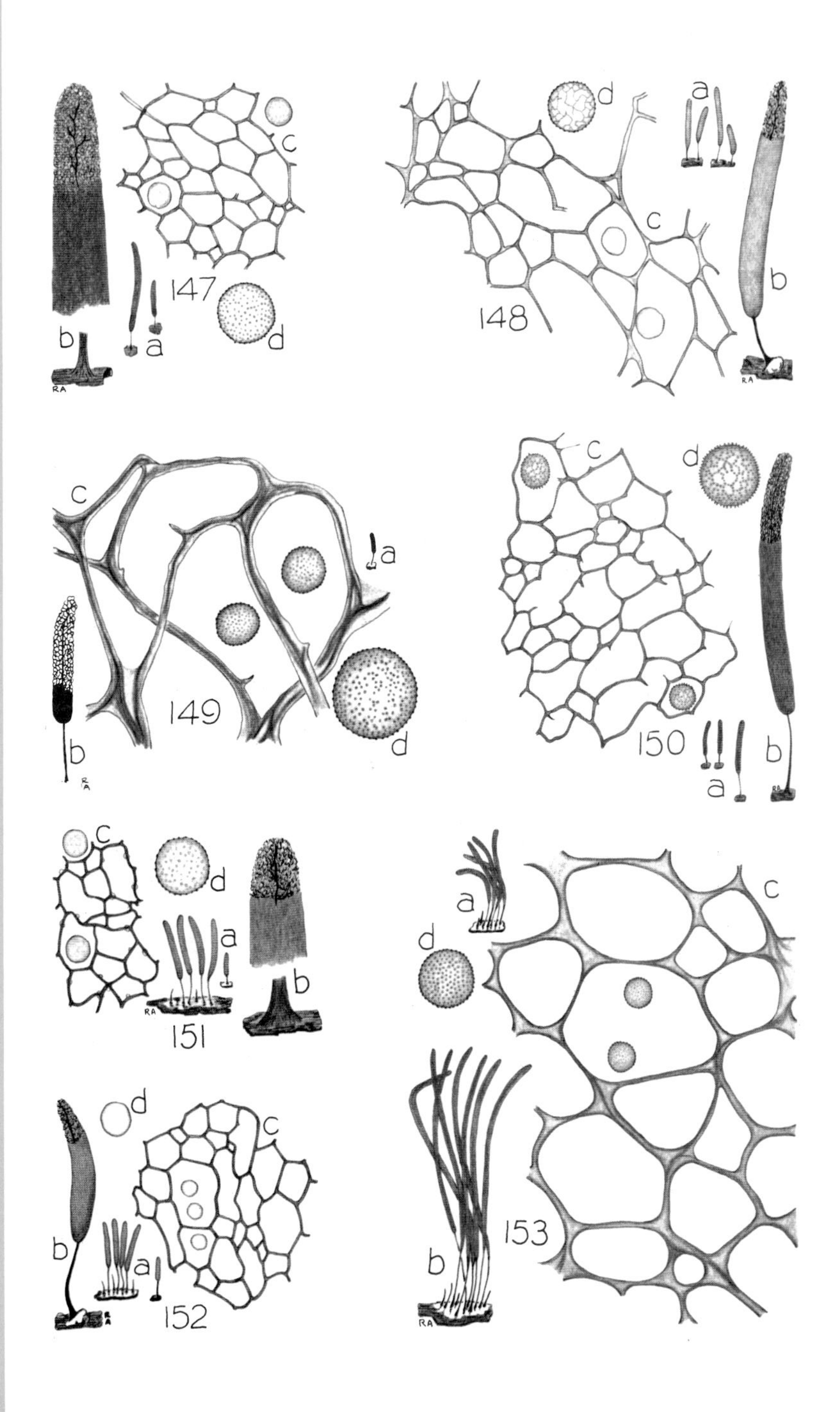

Plate XVII

154 *Stemonitis trechispora* (Berk.) Macbr.
- **a** Two sporangia, × 2
- **b** Sporangium, × 10
- **c** Detail of surface net, and spores, × 500
- **d** Spore, × 1000

155 *Stemonitis uvifera* Macbr.
- **a** Group of sporangia, × 2
- **b** Detail of surface net, and spores, × 500
- **c** Cluster of spores, × 1000

156 *Stemonitis virginiensis* Rex
- **a** Two sporangia, × 2
- **b** Sporangium from type, × 2
- **c** Same, × 10
- **d** Detail of surface net, and spores, × 500
- **e** Spore, × 1000

157 *Comatricha acanthodes* Alexop.
- **a** Sporangium, × 5
- **b** Same, spores fallen, × 50
- **c** Spore, × 1000

158 *Comatricha aequalis* Peck
- **a** Three sporangia, × 5
- **b** Detail of capillitium, and spores, × 500
- **c** Spore, × 1000

159 *Comatricha aggregata* Farr
- **a** Cluster of sporangia, × 5
- **b** Portion of same, lateral view, × 10
- **c** Isolated sporangium, × 20
- **d** Columella, showing enlarged tip, and attached capillitial threads and peridial fragments, × 50
- **e** Two spores, × 500
- **f** Spore, × 1000

160 *Comatricha caespitosa* Sturgis
- **a** Tuft of sporangia, × 5
- **b** Sporangium, × 20
- **c** Spore, × 1000

161 *Macbrideola decapillata* H. C. Gilbert
- **a** Three sporangia, × 5
- **b** Sporangium, spores fallen, × 50
- **c** Spore, × 1000. At the time this was drawn, it was believed to be *Comatricha cornea* G. Lister. See discussion under *Macbrideola.*

162 *Comatricha cylindrica* (Bilgram) Macbr.
- **a** Two sporangia, × 5
- **b** Sporangium, × 20
- **c** Spore, × 1000

163 *Comatricha elegans* (Racib.) G. Lister
- **a** Three sporangia, × 5
- **b** Blown sporangium, showing tip of stalk and characteristic branching leading to capillitium, × 50
- **c** Detail of capillitium, and spores, × 500
- **d** Spore, × 1000

164 *Comatricha fimbriata* G. Lister & Cran
- **a** Three sporangia, × 5
- **b** Sporangium, spores shed, × 50
- **c** Spore, × 1000

Plate XVII

Plate XVIII

165 *Comatricha irregularis* Rex
- **a** Three sporangia, showing range in size × 5
- **b** Median portion of sporangium, showing columella and capillitium, × 50
- **c** Detail of capillitium, and spores, × 500
- **d** Spore, × 1000

166 *Comatricha laxa* Rost.
- **a** Five sporangia, × 5
- **b** Sporangium, spores shed, × 20
- **c** Details of capillitium, and spores, × 500
- **d** Spore, × 1000

167 *Comatricha longa* Peck
- **a** Groups of sporangia from large fruitings, × ½
- **b** Section of sporangium, showing columella and capillitium, × 50
- **c** Detail of capillitium, and spores, × 250
- **d** Spore, × 1000

168 *Comatricha longipila* Nann.-Brem.
- **a** Three sporangia, × 5
- **b** Sporangium, × 20
- **c** Detail of outer portion of capillitium, with spores, × 500
- **d** Spore, × 1000

169 *Comatricha lurida* A. Lister
- **a** Four sporangia, × 5
- **b** Sporangium, showing stalk and capillitium, spores shed, × 50
- **c** Outer portion of capillitium, with spores, × 500
- **d** Spore, × 1000

170 *Macbrideola martinii* (Alexop. & Beneke) Alexop.
- **a** Two sporangia, × 5
- **b** Sporangium, showing stalk and capillitium, spores shed, × 50
- **c** Spore, × 1000

171 *Comatricha mirabilis* Benj. & Poit.
- **a** Three sporangia, × 5
- **b** Sporangium, showing stalk and capillitium, × 20
- **c** Tips of capillitium and spores, × 250
- **d** Spore, × 1000

172 *Comatricha nigra* (Pers.) Schroet.
- **a** Three sporangia, × 5
- **b** Sporangium, spores shed, × 20
- **c** Capillitium and spores, × 500
- **d** Spore, × 1000

173 *Comatricha pulchella* (C. Bab.) Rost.
- **a** Three sporangia, × 5
- **b** Sporangium, spores shed, × 20
- **c** Capillitium at margin, and spores, × 500
- **d** Spore, × 1000

Plate XVIII

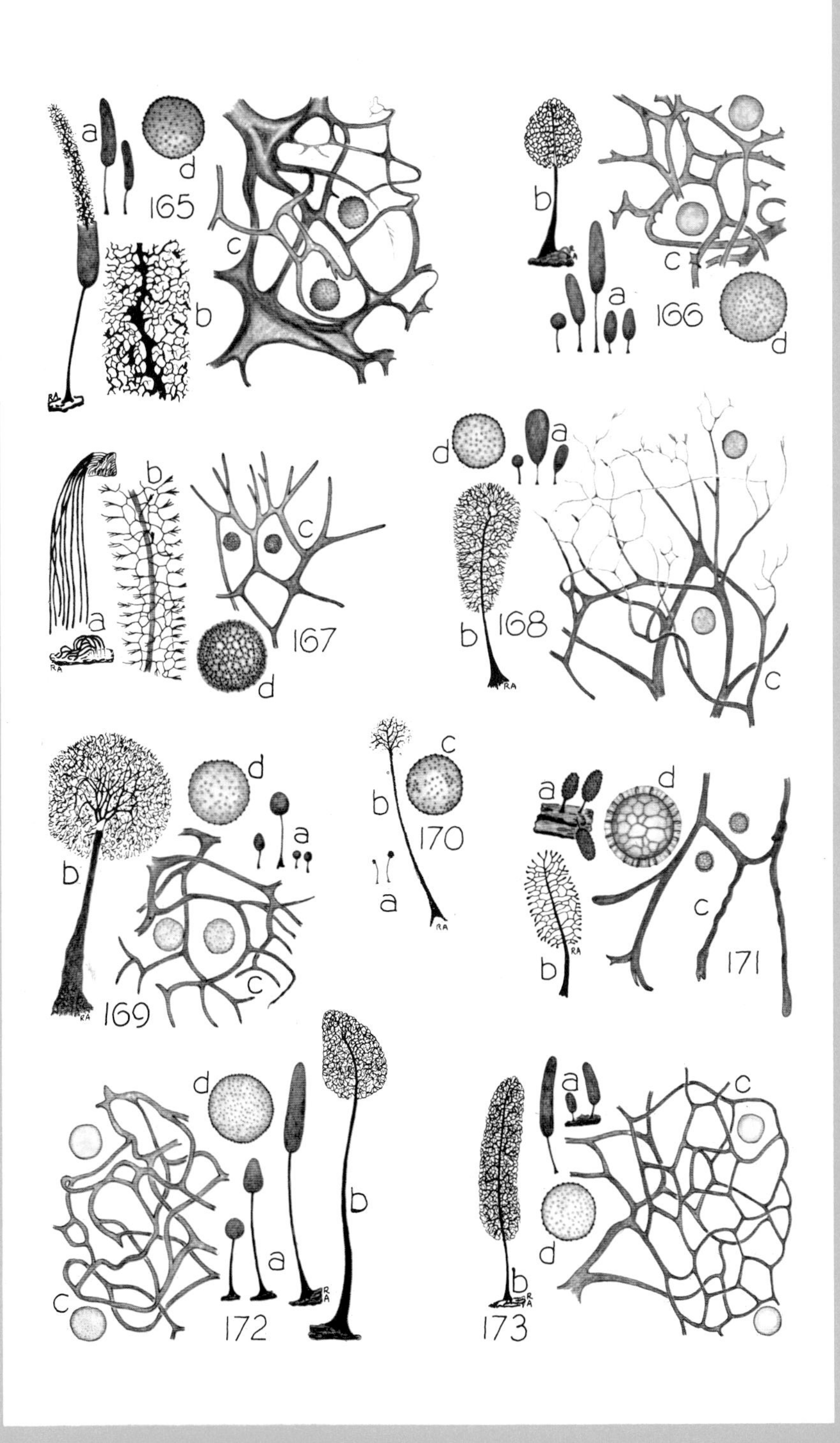

Plate XIX

174 *Comatricha dictyospora* Čelak.
a Three sporangia, × 5
b Sporangium, × 10
c Detail of capillitium, and spores, × 500
d Spore, × 1000. Drawn from type of *C. reticulata* H. C. Gilbert

175 *Comatricha rispaudii* Hagelst.
a Tuft of sporangia, × 5
b Sporangium, showing persistent peridium at base, × 20
c Detail of capillitium, and spores, × 500
d Spore, × 1000

176 *Comatricha rubens* A. Lister
a Three sporangia, × 5
b Sporangium, × 20
c Tips of capillitium, and spores, × 500
d Spore, × 1000

177 *Comatricha subcaespitosa* Peck
a Three sporangia, × 5
b Sporangium, × 20
c Margin of capillitium, and spores, × 500
d Spore, × 1000

178 *Comatricha suksdorfii* Ell. & Ev.
a Three sporangia, × 5
b Detail of capillitium, and spores, × 500
c Spore, × 1000

179 *Macbrideola synsporos* (Alexop.) Alexop.
a Two sporangia, × 5
b Sporangium, with spore clusters, as seen in mounted specimen, × 100
c Cluster of spores, × 500
d Isolated spore, × 1000

180 *Comatricha tenerrima* (M. A. Curt.) G. Lister
a Three sporangia, × 5
b Sporangium, × 10
c Detail of capillitium, and spores, × 500
d Spore, × 1000

181 *Comatricha typhoides* (Bull.) Rost.
a Three sporangia, × 5
b Sporangium, × 10
c Detail of capillitium, and spores, × 500
d Spore, × 1000

Plate XIX

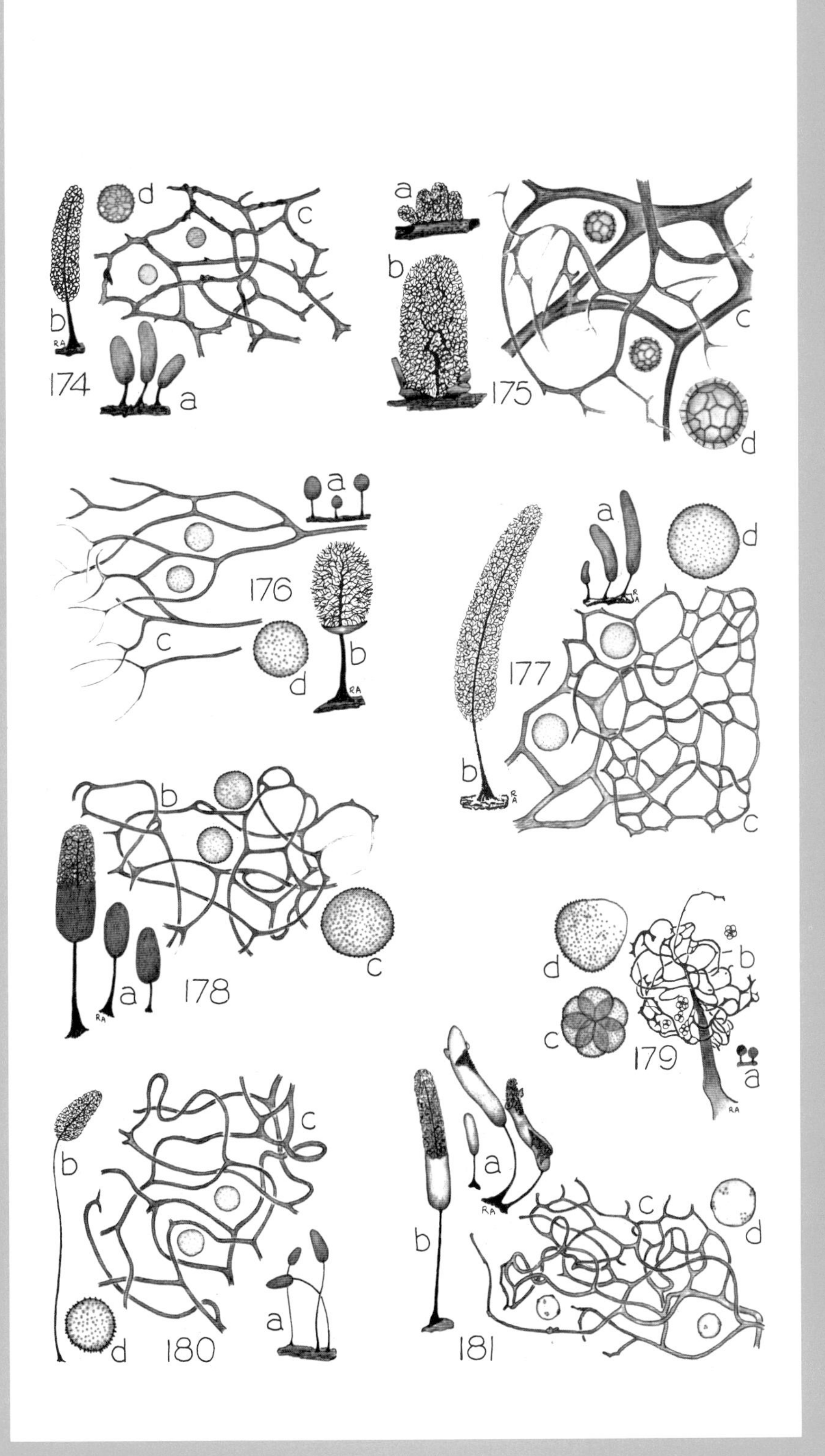

174
a
b
c
d
175
a
b
c
d
176
a
b
c
d
177
a
b
c
d
178
a
b
c
179
a
b
c
d
180
a
b
c
d
181
a
b
c
d

Plate XX

182 *Macbrideola decapillata* H. C. Gilbert
- **a** Two sporangia, × 30
- **b** Sporangium bearing spores, × 150
- **c** Same, showing columella, × 150
- **d** Spore, × 1000

183 *Macbrideola scintillans* H. C. Gilbert
- **a** Two sporangia, × 30
- **b** Sporangium, showing capillitium, spores and persistent peridium, × 150
- **c** Spore, × 1000

184 *Clastoderma debaryanum* Blytt
- **a** Three sporangia, × 5
- **b** Two sporangia, one with spores shed, × 30
- **c** Sporangium, with capillitium, spores and upper part of stalk, × 100
- **d** Tips of capillitium, with attached scales, and spores, × 500
- **e** Spore, × 1000

185 *Barbeyella minutissima* Meylan
- **a** Two sporangia, × 5
- **b** Sporangium, × 50
- **c** Capillitium arising from columella, and collar at base of sporangium, × 200
- **d** Spore, × 1000

186 *Lamproderma arcyrioides* (Sommerf.) Rost.
- **a** Four sporangia, illustrating variation, × 5
- **b** Two sporangia, showing peridium and capillitium, × 20
- **c** Tips of capillitium, and spores, × 500
- **d** Spore, × 1000

187 *Lamproderma arcyrionema* Rost.
- **a** Three sporangia, showing variation, × 5
- **b** Two sporangia, × 20
- **c** Tips of capillitium, and spores, × 500
- **d** Spore, × 1000

188 *Lamproderma carestiae* (Ces. & de Not.) Meylan
- **a** Stalked and sessile sporangium, × 5
- **b** Two sporangia, × 15
- **c** Tips of capillitium, and spores, × 500
- **d** Spore, × 1000

189 *Lamproderma columbinum* (Pers.) Rost.
- **a** Two sporangia, × 5
- **b** Two sporangia, × 10
- **c** Tips of capillitium, and spores, × 500
- **d** Spore, × 1000

190 *Lamproderma cribrarioides* (Fries) R. E. Fries
- **a** Two sporangia, × 5
- **b** Two sporangia, × 10
- **c** Tips of capillitium, and spores, × 500
- **d** Spore, × 1000

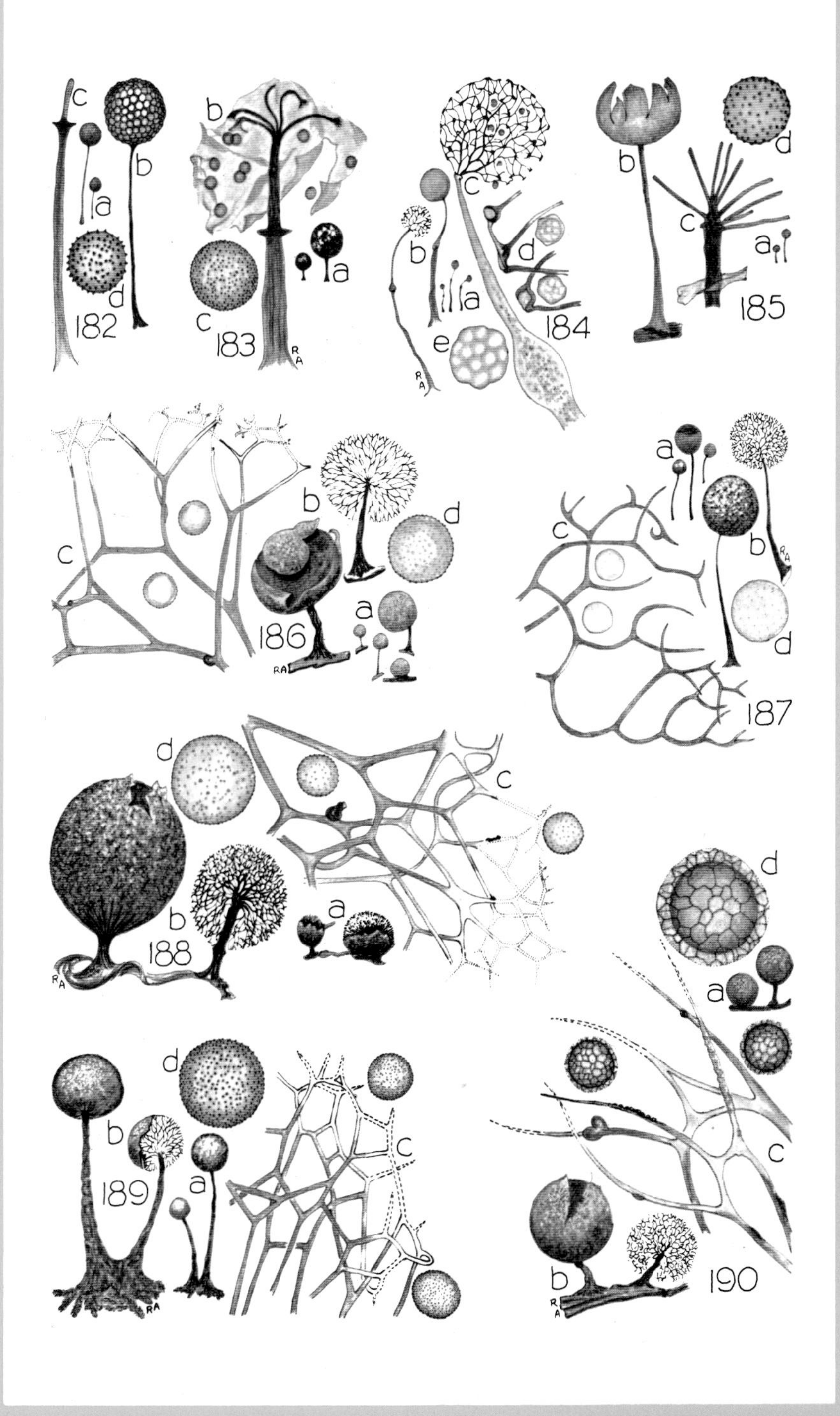

c
b
a
d
182
b
a
c
183
c
b
a
d
e
184
b
d
c
a
185
c
b
d
a
186
a
c
b
d
187
d
c
b
a
188
d
b
a
c
189
d
a
c
b
190

Plate XXI

191 *Lamproderma cristatum* Meylan
- **a** Two sporangia, × 5
- **b** Sporangium, showing columella, capillitium and persistent peridium at base, × 10
- **c** Spore, × 1000

192 *Lamproderma echinulatum* (Berk.) Rost.
- **a** Two sporangia, × 5
- **b** Obovate sporangium, × 10
- **c** Spore, × 1000

193 *Lamproderma gulielmae* Meylan
- **a** Two sporangia, × 5
- **b** Two sporangia, × 20
- **c** Spore, × 1000

194 *Lamproderma muscorum* (Lév.) Hagelst.
- **a** Two sporangia, × 5
- **b** Sporangium, × 50
- **c** Spore, × 1000

195 *Lamproderma pulchellum* Meylan
- **a** Cluster of sporangia, × 5
- **b** Two sporangia, × 20
- **c** Denuded sporangium, × 20
- **d** Detail of capillitium, and spores, × 500
- **e** Spore, × 1000

196 *Lamproderma atrosporum* Meylan
- **a** Four sporangia, × 5
- **b** Denuded sporangium, × 20
- **c** Detail of capillitium showing tips attached to fragments of peridium, and spores, × 500
- **d** Spore, × 1000

197 *Lamproderma sauteri* Rost.
- **a** Group of sporangia, × 5
- **b** Sporangium, × 15
- **c** Tips of capillitium, and spores, × 500
- **d** Spore, × 1000

198 *Lamproderma scintillans* (Berk. & Br.) Morgan
- **a** Three sporangia, × 5
- **b** Sporangium, × 40
- **c** Capillitium, columella and tip of stalk, × 40
- **d** Capillitium arising from tip of columella, showing pale bases of threads, × 100
- **e** Spore, × 1000

199 *Lamproderma verrucosum* Martin, Thind & Sohi
- **a** Two sporangia, × 5
- **b** Two sporangia, × 50
- **c** Tips of capillitium, and spores, × 500
- **d** Spore, × 1000

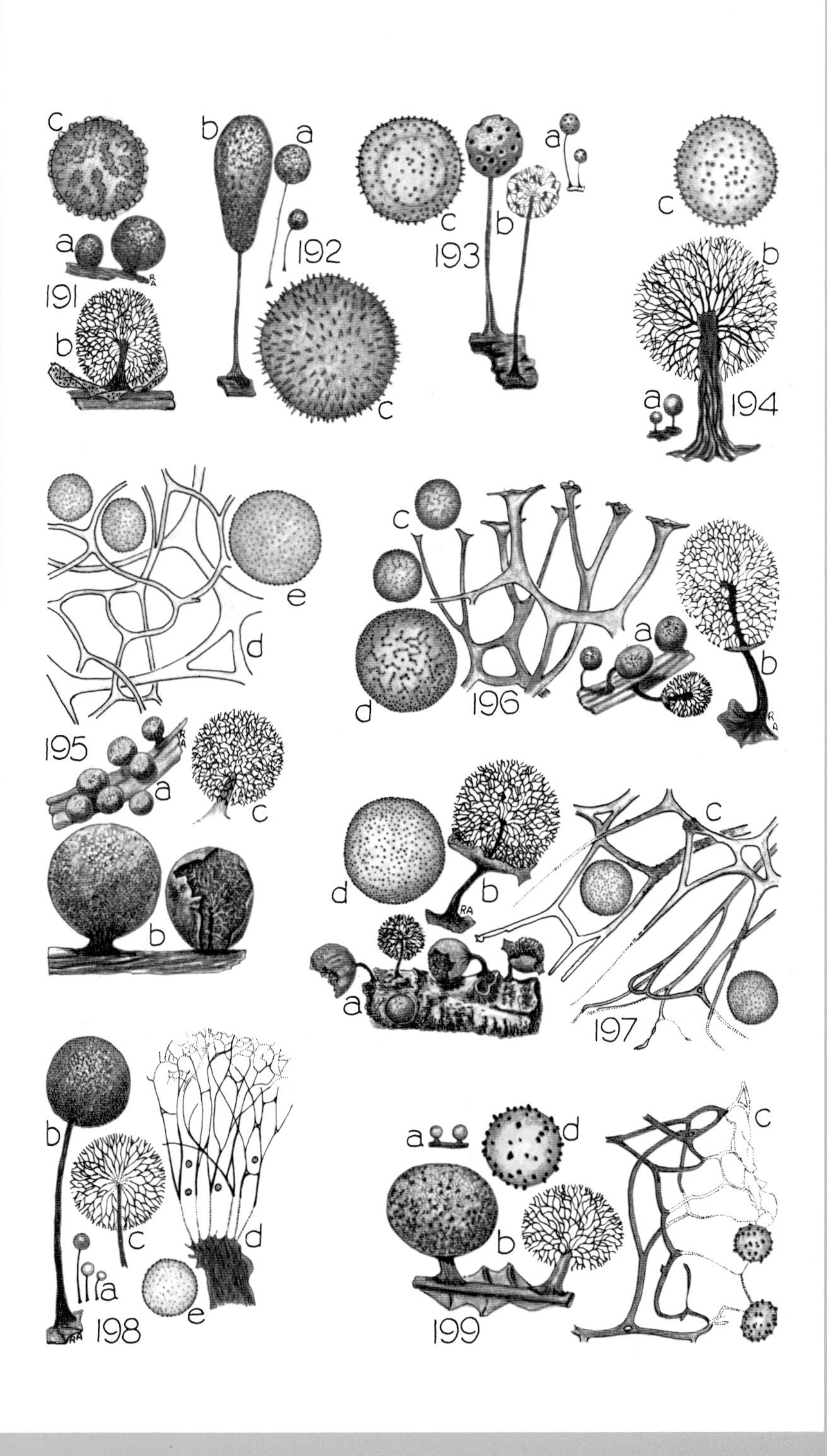
191
a
b
c
192
a
b
c
193
a
b
c
194
a
b
c
195
a
b
c
d
e
196
a
b
c
d
197
a
b
c
d
198
a
b
c
d
e
199
a
b
c
d

Plate XXII

200 *Cienkowskia reticulata* (Alb. & Schw.) Rost.
- **a** Plasmodiocarp, × 10
- **b** Portion of slender plasmodiocarp, × 5
- **c** Detail of plasmodiocarp, showing limy plates, × 50
- **d** Detail of capillitium, and spores, × 250
- **e** Spore, × 1000

201 *Leocarpus fragilis* (Dicks.) Rost.
- **a** Sporangia, × 5
- **b** Broken sporangium, × 20
- **c** Detail of capillitium, and spores, × 250
- **d** Spore, × 1000

202 *Physarella oblonga* (Berk. & Curt.) Morgan
- **a** Three sporangia, × 5
- **b** Open sporangium, showing spikes and pseudo-columella, × 20
- **c** Sporangium before dehiscence, × 20
- **d** Capillitium, spike, and spores, × 250
- **e** Spore, × 1000

203 *Badhamia affinis* Rost.
- **a** Cluster of sporangia, × 5
- **b** Open sporangium, × 20
- **c** Capillitium attached to peridium, and spores, × 250
- **d** Spore, × 1000

204 *Badhamia capsulifera* (Bull.) Berk.
- **a** Group of sporangia, × 5
- **b** Single sporangium, × 20
- **c** Cluster of spores, × 500
- **d** Isolated spore, × 1000

205 *Badhamia dearnessii* Hagelst.
- **a** Group of sporangia, × 5
- **b** Single sporangium, × 20
- **c** Two spores, × 1000

206 *Badhamia foliicola* A. Lister
- **a** Cluster of sporangia, × 5
- **b** Sporangium, × 20
- **c** Spore, × 1000

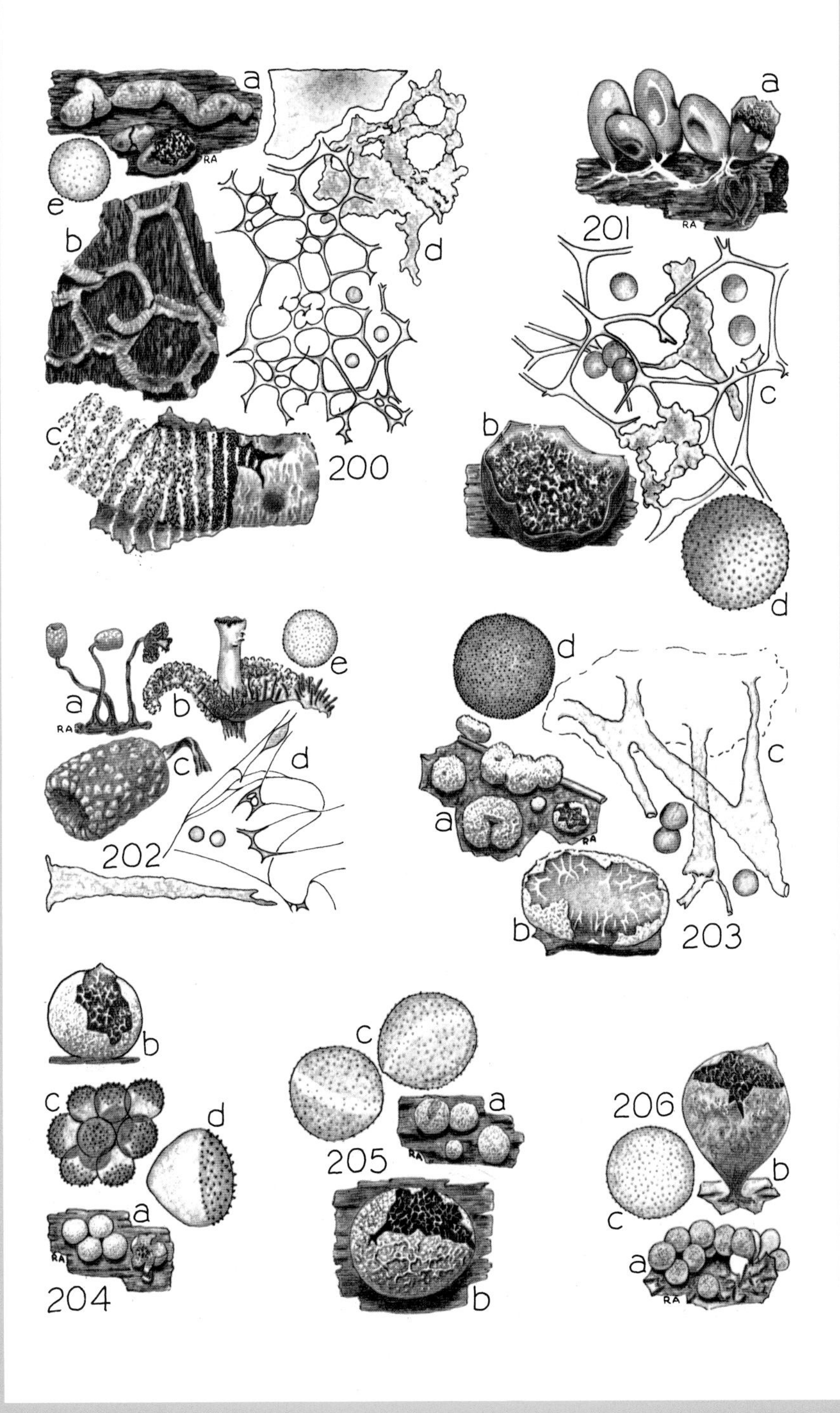
a
e
b
d
c
200
RA
a
201
RA
c
b
d
a
RA
b
e
c
d
202
d
c
a
RA
b
203
b
c
d
a
RA
204
c
a
205
RA
b
206
b
c
a
RA

Plate XXIII

207 *Badhamia gracilis* (Macbr.) Macbr.
- **a** Group of sporangia, × 5
- **b** Sporangium, × 20
- **c** Two spores, × 1000

208 *Badhamia lilacina* (Fries) Rost.
- **a** Group of sporangia, × 5
- **b** Two sporangia, × 20
- **c** Open sporangium, × 20
- **d** Spore, × 1000

209 *Badhamia macrocarpa* (Ces.) Rost.
- **a** Group of sporangia, × 5
- **b** Stalked sporangium, × 20
- **c** Detail of capillitium, and spores, × 250
- **d** spore, × 1000

210 *Badhamia nitens* Berk.
- **a** Cluster of sporangia (after Lister), × 5
- **b** Sporangium, × 20
- **c** Cluster of spores, × 500
- **d** Same, × 1000

211 *Badhamia obovata* (Peck) S. J. Smith
- **a** Cluster of sporangia (Iowa), × 5
- **b** Same (Massachusetts), × 5
- **c** Sporangium of *B. rubiginosa* var. *globosa* (Wales), × 5
- **d** Two sporangia (Iowa), one showing columella, × 20
- **e** Spore (Iowa), × 1000
- **f** Same of var. *dictyospora,* × 1000
- **g** Same of var. *globosa* (Wales), × 1000

212 *Badhamia ovispora* Racib.
- **a** Cluster of sporangia and plasmodiocarps, × 5
- **b** Subplasmodiocarpous sporangium, × 50
- **c** Spore, × 1000

213 *Badhamia panicea* (Fries) Rost.
- **a** Group of sporangia, × 5
- **b** Sporangium, × 20
- **c** Two spores, × 500
- **d** Spore, × 1000

214 *Badhamia papaveracea* Berk. & Rav.
- **a** Three sporangia, × 5
- **b** Sporangium, × 20
- **c** Cluster of spores, with free spore, × 500
- **d** Spore ball, × 1000

215 *Badhamia populina* A. & G. Lister
- **a** Cluster of sporangia, × 5
- **b** Sporangium, × 10
- **c** Spores, partly clustered, × 500
- **d** Spore, × 1000

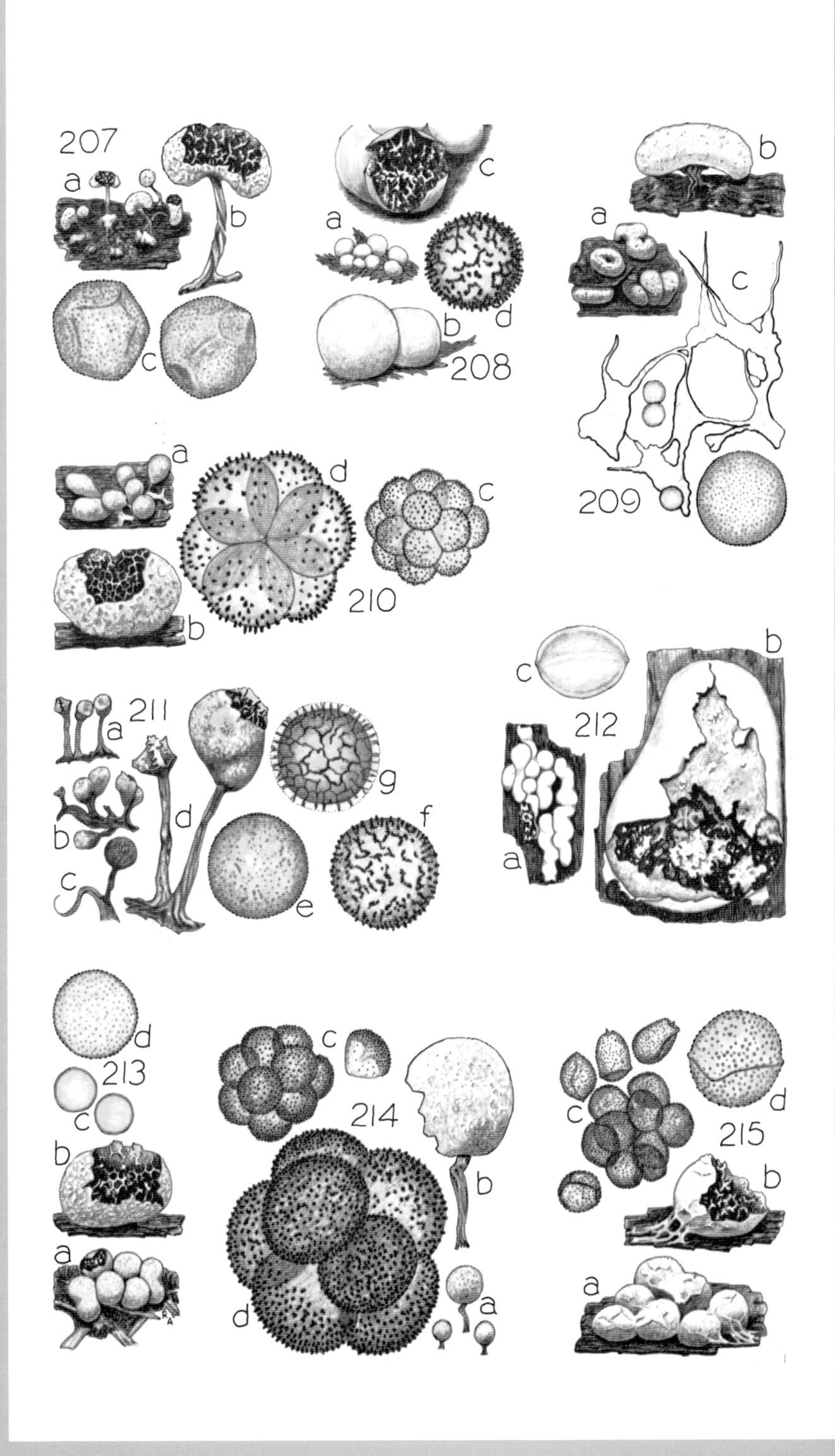
207
a
b
c
208
a
b
c
d
209
a
b
c
210
a
b
c
d
211
a
b
c
d
e
f
g
212
a
b
c
213
a
b
c
d
214
a
b
c
d
215
a
b
c
d

Plate XXIV

216 *Badhamia utricularis* (Bull.) Berk.
  a Cluster of pendent sporangia, × 5
  b Loose cluster of spores, × 500
  c Spore, × 1000

217 *Badhamia versicolor* A. Lister
  a Three sporangia, × 5
  b Sporangium, peridium partly shed, × 20
  c Cluster of spores and two free spores, × 500
  d Spore, × 1000

218 *Badhamia viridescens* Meylan
  a Two sporangia, × 5
  b Sporangium, × 20
  c Two spores, × 500
  d Spore, × 1000

219 *Fuligo cinerea* (Schw.) Morgan
  a Fragment of reticulate aethalium, × 1
  b Detail of same, showing capillitium imbedded in spore-mass, × 20
  c Detail of capillitium, and spores, × 250
  d Spore, × 1000

220 *Fuligo intermedia* Macbr.
  a Small aethalium, × 1
  b Detail of broken aethalium, × 20
  c Detail of capillitium, and spores, × 250
  d Spore, × 1000

221 *Fuligo megaspora* Sturgis
  a Aethalium, × 1
  b Detail of broken aethalium, × 20
  c Detail of capillitium, and spores, × 250
  d Spore, × 1000

222 *Fuligo muscorum* Alb. & Schw.
  a Aethalium, × 1
  b Detail of same, × 20
  c Capillitium and spores, × 250
  d Spore, × 1000

223 *Fuligo septica* (L.) Wiggers
  a Small aethalium, × 1
  b Same, in section, × 1
  c Detail with cortex removed, × 20
  d Detail of capillitium, and spores, × 250
  e Spore, × 1000

224 *Erionema aureum* Penzig
  a Pendent plasmodiocarps, × 10
  b Detail, × 25
  c Detail of capillitium, and spores, × 250
  d Spore, × 1000

Plate XXIV

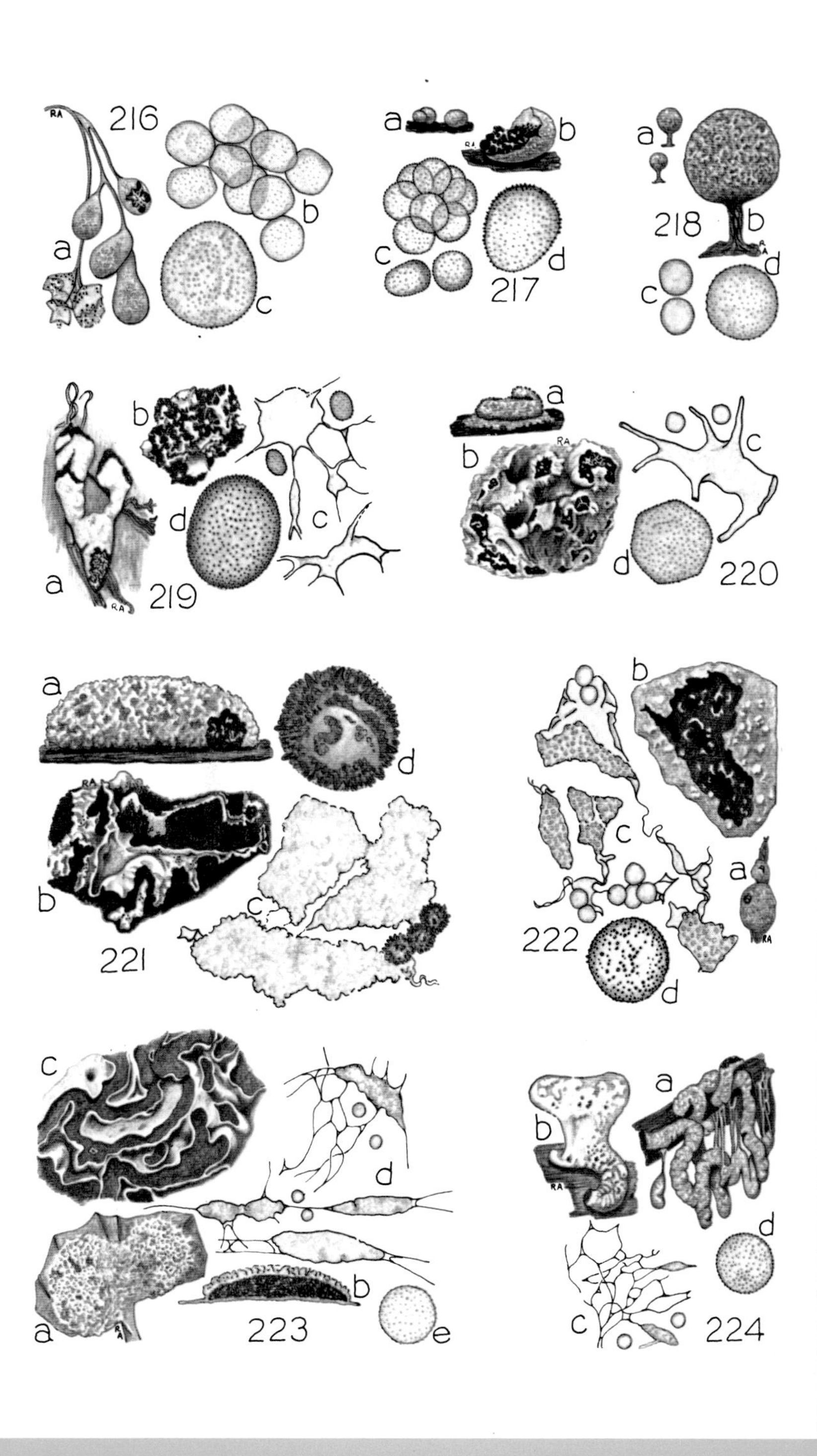
216
a
b
c
217
a
b
c
d
218
a
b
c
d
219
a
b
c
d
220
a
b
c
d
221
a
b
c
d
222
a
b
c
d
223
a
b
c
d
e
224
a
b
c
d

Plate XXV

225 *Craterium aureum* (Schum.) Rost.
- **a** Four sporangia, × 5
- **b** Three sporangia, × 20
- **c** Detail of capillitium, and spores, × 250
- **d** Spore, × 1000

226 *Craterium concinnum* Rex
- **a** Three sporangia, × 5
- **b** Two sporangia, × 20
- **c** Sporangium, × 50
- **d** Detail of capillitium, and spores, × 250
- **e** Spore, × 1000

227 *Craterium leucocephalum* (Pers.) Ditmar
- **a** Six sporangia, to show variation, × 5
- **b** Two sporangia, × 20
- **c** Detail of capillitium, and spores, × 250
- **d** Spore, × 1000

228 *Craterium minutum* (Leers) Fries
- **a** Four sporangia, × 5
- **b** Two sporangia, × 20
- **c** Detail of capillitium, and spores, × 250
- **d** Spore, × 1000

229 *Craterium paraguayense* (Speg.) G. Lister
- **a** Four sporangia, × 5
- **b** Two sporangia, × 20
- **c** Detail of capillitium, and spores, × 250
- **d** Spore, × 1000

230 *Physarum aeneum* (A. Lister) R. E. Fries
- **a** Sporangium and small plasmodiocarp, × 5
- **b** Portion of plasmodiocarp, × 20
- **c** Detail of capillitium, and spores, × 250
- **d** Spore, × 1000

231 *Physarum albescens* Ellis
- **a** Group of sporangia, × 5
- **b** Sporangium, × 20
- **c** Detail of capillitium, and spores, × 250
- **d** Spore, × 1000

232 *Physarum alpinum* (A. & G. Lister) G. Lister
- **a** Cluster of fructifications, × 5
- **b** Portion of same, × 20
- **c** Detail of capillitium, and spores, × 250
- **d** Spore, × 1000

233 *Physarum auripigmentum* Martin
- **a** Two sporangia, × 5
- **b** Two sporangia, × 20
- **c** Detail of capillitium, and spores, × 250
- **d** Spore, × 1000

Plate XXV

225
a
b
c
d
226
a
b
c
d
e
227
a
b
c
d
228
a
b
c
d
229
a
b
c
d
230
a
b
c
d
231
a
b
c
d
232
a
b
c
d
233
a
b
c
d

Plate XXVI

234 *Physarum auriscalpium* Cooke
- **a** Two plasmodiocarps and pulvinate sporangium, yellow phase, × 5
- **b** Plasmodiocarp, green phase, × 5
- **c** Pulvinate sporangium, with limeless base, × 20
- **d** Detail of capillitium, and spores, × 250
- **e** Spore, × 1000

235 *Physarum bethelii* Macbr.
- **a** Two sporangia, × 5
- **b** Sporangium, × 20
- **c** Nearly empty sporangium (type), × 20
- **d** Detail of capillitium, and spores, × 250
- **e** Spore, × 1000

236 *Physarum bilgramii* Hagelst.
- **a** Two sporangia, different collections, × 5
- **b** Sporangium, and tip of stalk, showing columella, × 20
- **c** Two sporangia, from other fruitings, × 20
- **d** Detail of capillitium, and spores, × 250
- **e** Spore, × 1000

237 *Physarum bitectum* G. Lister
- **a** Cluster of fructifications, × 5
- **b** Sporangium, × 20
- **c** Detail of capililitium, and spores, × 250
- **d** Spore, × 1000

238 *Physarum bivalve* Pers.
- **a** Cluster of fructifications, × 5
- **b** Plasmodiocarp, × 10
- **c** Detail of capillitium, and spores, × 250
- **d** Spore, × 1000

239 *Physarum bogoriense* Racib.
- **a** Portion of plasmodiocarp, × 5
- **b** Same, × 20
- **c** Sporangiate fruiting, × 20
- **d** Detail of capillitium, and spores, × 250
- **e** Spore, × 1000

240 *Physarum braunianum* de Bary
- **a** Five sporangia, × 5
- **b** Sporangium, × 50
- **c** Detail of capillitium, and spores, × 250
- **d** Spore, × 1000

241 *Physarum brunneolum* (Phill.) Massee
- **a** Sporangium, × 5
- **b** Sporangium, × 20
- **c** Same, from above showing double wall, × 20
- **d** Detail of capillitium, and spores, × 250
- **e** Spore, × 1000

242 *Physarum carneum* G. Lister & Sturgis
- **a** Three sporangia, × 5
- **b** Two sporangia, × 20
- **c** Detail of capillitium, and spores, × 250
- **d** Two spores, × 1000

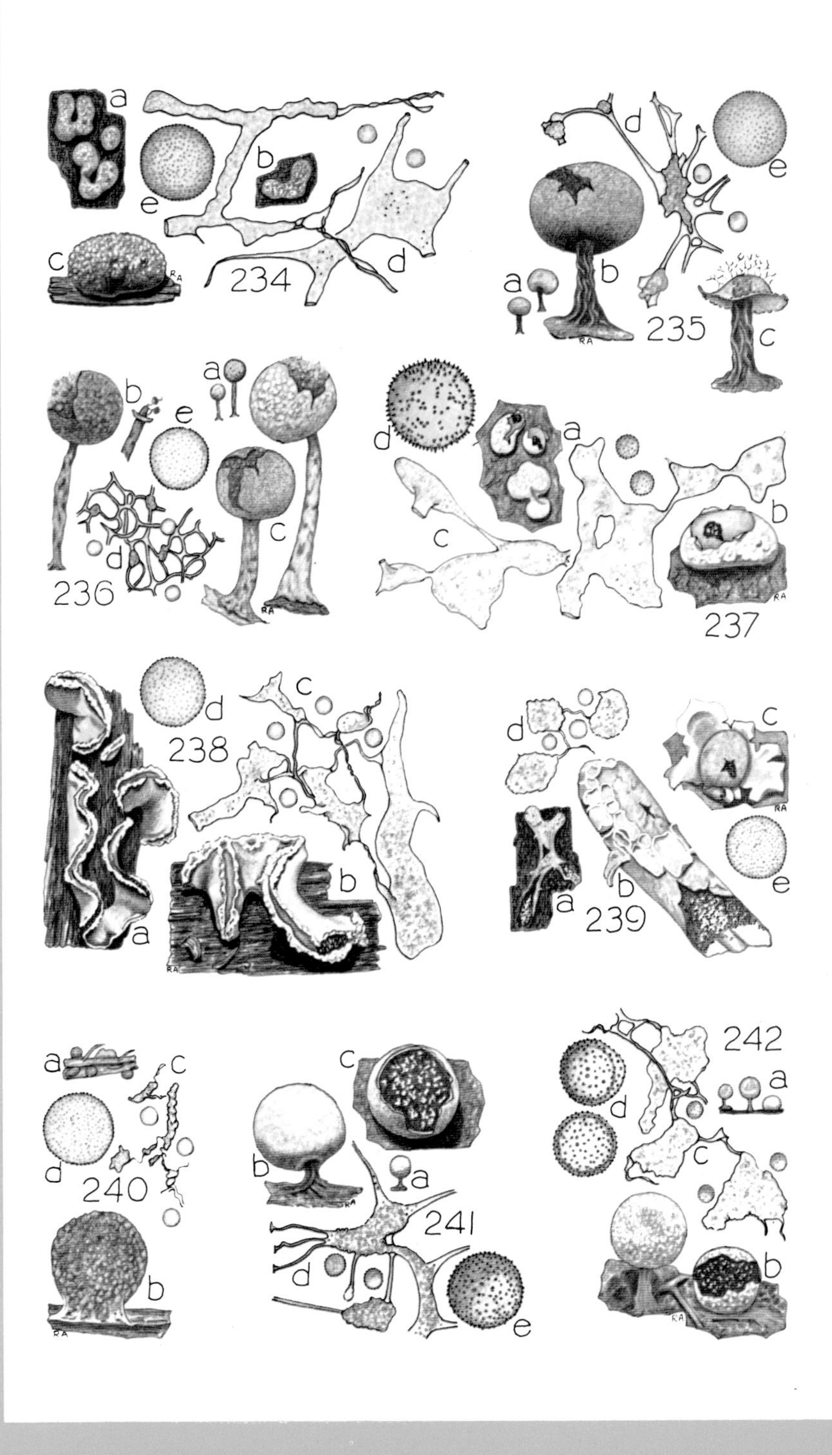

234
a
b
c
d
e
235
a
b
c
d
e
236
a
b
c
d
e
237
a
b
c
d
238
a
b
c
d
239
a
b
c
d
e
240
a
b
c
d
241
a
b
c
d
e
242
a
b
c
d

Plate XXVII

243 *Physarum cinereum* (Batsch) Pers.
- **a** Cluster of sporangia, × 5
- **b** Smaller cluster, showing subplasmodiocarpous fruitings, × 20
- **c** Detail of capillitium, and spores, × 250
- **d** Spore, × 1000

244 *Physarum citrinum* Schum.
- **a** Three sporangia, × 5
- **b/c** Two sporangia, × 20
- **d** Detail of capillitium, and spores, × 250
- **e** Spore, × 1000

245 *Physarum compressum* Alb. & Schw.
- **a** Two sporangia, × 5
- **b** Sporangium and stalk of another, × 20
- **c** Detail of capillitium, and spores, × 250
- **d** Spore, × 1000

246 *Physarum confertum* Macbr.
- **a** Cluster of sporangia, × 5
- **b** Same, × 20
- **c** Detail of capillitium, and spores, × 250
- **d** Spore, × 1000

247 *Physarum contextum* (Pers.) Pers.
- **a** Portion of an extensive fruiting, × 5
- **b** Same, × 20
- **c** Detail of capillitium, and spores, × 250
- **d** Spore, × 1000

248 *Physarum crateriforme* Petch
- **a** Two sporangia, × 5
- **b** Two sporangia, the one at right showing columella, × 20
- **c** Detail of capillitium, and spores, × 250
- **d** Spore, × 1000

249 *Physarum decipiens* Curtis
- **a** Cluster of sporangia, × 5
- **b** Same, × 20
- **c** Detail of capillitium, and spores, × 250
- **d** Spore, × 1000

250 *Physarum dictyosporum* Martin
- **a** Three sporangia, × 5
- **b** Three fructifications approaching plasmodiocarp type, × 20
- **c** Detail of capillitium, and spores, × 250
- **d** Spore, × 1000

Plate XXVII

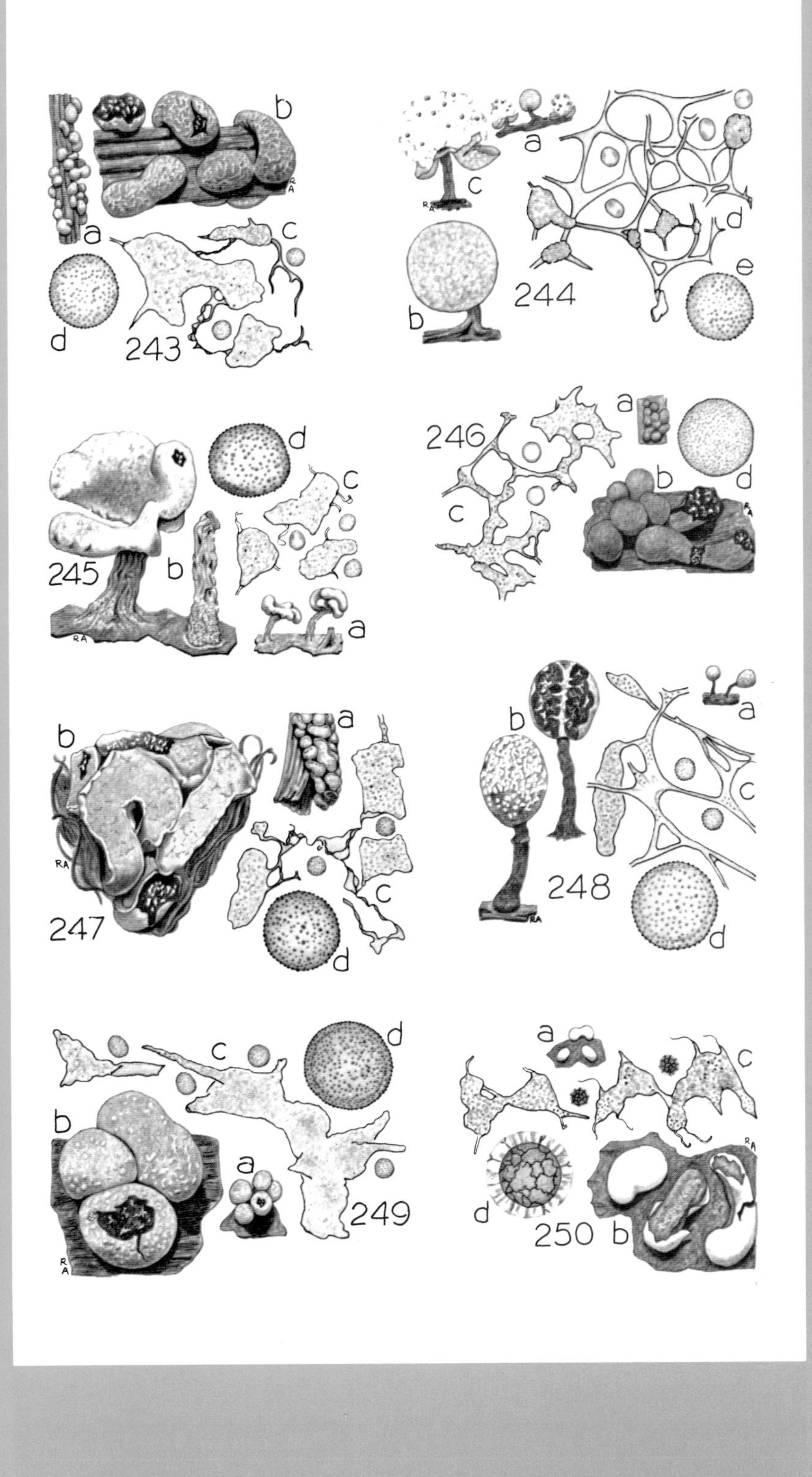

243
a
b
c
d
244
a
b
c
d
e
245
a
b
c
d
246
a
b
c
d
247
a
b
c
d
248
a
b
c
d
249
a
b
c
d
250
a
b
c
d

Plate XXVIII

251 *Physarum diderma* Rost.
- **a** Cluster of sporangia, × 5
- **b** Sporangium, × 20
- **c** Detail of capillitium, and spores, × 250
- **d** Spore, × 1000

252 *Physarum didermoides* (Pers.) Rost.
- **a** Portion of large fruiting, × 5
- **b** Three sporangia, × 20
- **c** Detail of capillitium, and spores, × 250
- **d** Spore, × 1000

253 *Physarum digitatum* G. Lister & Farq.
- **a** Cluster of sporangia, × 5
- **b** Detail of same, × 20
- **c** Detail of capillitium, and spores, × 250
- **d** Spore, × 1000

254 *Physarum echinosporum* G. Lister
- **a** Cluster of plasmodiocarps, × 5
- **b** Two plasmodiocarps, × 10
- **c** Detail of capillitium, and spores, × 250
- **d** Spore, × 1000

255 *Physarum flavicomum* Berk.
- **a** Three sporangia with stalk of another, × 5
- **b** Two sporangia, × 20
- **c** Detail of capillitium, and spores, × 250
- **d** Spore, × 1000

256 *Physarum flavidum* (Peck) Peck
- **a** Four sporangia, × 5
- **b** Sporangium, × 20
- **c** Detail of capillitium, and spores, × 250
- **d** Spore, × 1000

257 *Physarum galbeum* Wingate
- **a** Three sporangia, × 5
- **b** Two sporangia, × 20
- **c** Detail of capillitium, and spores, × 250
- **d** Spore, × 1000

258 *Physarum globuliferum* (Bull.) Pers.
- **a** Two sporangia, with stalks of two others, one showing columella, × 5
- **b** Other sporangia and stalks, to show variation, × 5
- **c** Two sporangia, with tip of another showing columella, × 20
- **d** Sporangium, pinkish phase with scanty lime, × 20
- **e** Detail of capillitium, and spores, × 250
- **f** Spore, × 1000

259 *Physarum gyrosum* Rost.
- **a** Small plasmodiocarp, × 5
- **b** Same, × 20
- **c** Detail of same, showing spike-like nodes, × 50
- **d** Detail of capillitium, and spores, × 250
- **e** Spore, × 1000

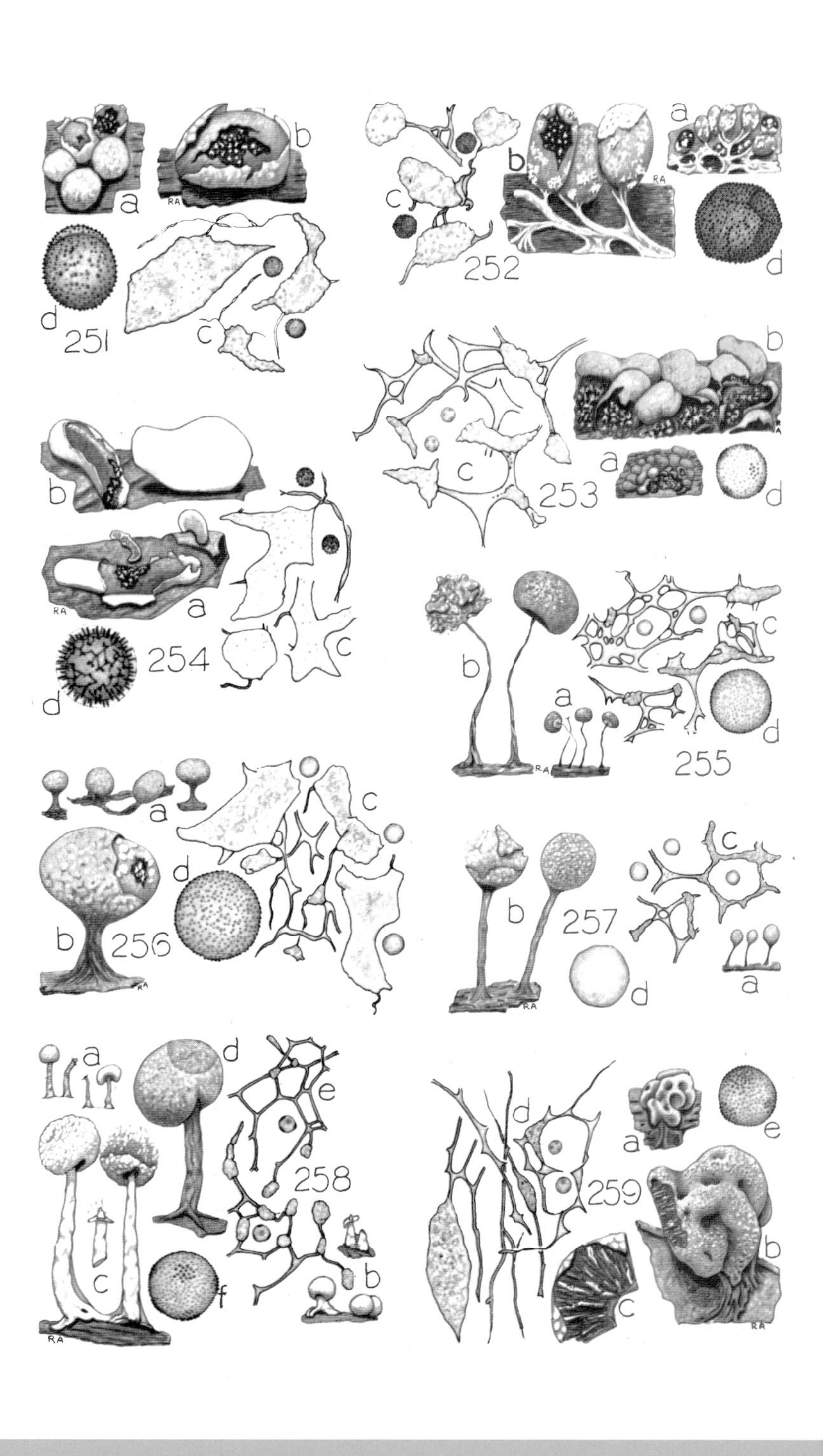
251
a
b
c
d
252
a
b
c
d
253
a
b
c
d
254
a
b
c
d
255
a
b
c
d
256
a
b
c
d
257
a
b
c
d
258
a
b
c
d
e
f
259
a
b
c
d
e

Plate XXIX

260 *Physarum javanicum* Racib.
- **a** Cluster of sporangia, × 5
- **b** Sporangium, × 20
- **c** Detail of capillitium, and spores, × 250
- **d** Spore, × 1000

261 *Physarum lateritium* (Berk. & Rav.) Morgan
- **a** Two sporangia, × 5
- **b** Sporangium, × 20
- **c** Detail of capillitium, and spores, × 250
- **d** Spore, × 1000

262 *Physarum leucophaeum* Fries
- **a** Three sporangia, × 5
- **b** Two sporangia, × 20
- **c** Detail of capillitium, and spores, × 250
- **d** Spore, × 1000

263 *Physarum leucopus* Link
- **a** Sporangium, × 5
- **b** Two sporangia, × 20
- **c** Detail of capillitium, and spores, × 250
- **d** Spore, × 1000

264 *Physarum listeri* Macbr.
- **a** Two sporangia, × 5
- **b** Two sporangia, × 10
- **c** Portion of stem, showing crystalline inclusions, × 50
- **d** Detail of capillitium, and spores, × 250
- **e** Spore, × 1000

265 *Physarum lutescens* Peck
- **a** Cluster of sporangia, × 5
- **b** Sporangium, × 20
- **c** Detail of capillitium, and spores, × 250
- **d** Spore, × 1000

266 *Physarum megalosporum* Macbr.
- **a** Three sporangia, × 5
- **b** Two sporangia, × 20
- **c** Detail of capillitium, and spores, × 250
- **d** Spore, × 1000

267 *Physarum melleum* (Berk. & Br.) Massee
- **a** Four sporangia, × 5
- **b** Sporangium, and base of another, showing columella, × 20
- **c** Detail of capillitium, and spores, × 250
- **d** Spore, × 1000

268 *Physarum mennegae* Nann.-Brem.
- **a** Sporangium, × 5
- **b** Same, with stalk showing columella, × 20
- **c** Detail of capillitium, and spores, × 250
- **d** Spore, × 1000

269 *Physarum mortoni* Macbr.
- **a** Cluster of sporangia, × 5
- **b** Sporangium, with weak, prostrate stalk, × 20
- **c** Detail of capillitium, and spores, × 250
- **d** Spore, × 1000

260
a
b
c
d
261
a
b
c
d
262
a
b
c
d
263
a
b
c
d
264
a
b
c
d
e
265
a
b
c
d
266
a
b
c
d
267
a
b
c
d
268
a
b
c
d
269
a
b
c
d

Plate XXX

270 *Physarum murinum* A. Lister
- **a** Two sporangia, × 5
- **b** Two sporangia, one showing columella, × 20
- **c** Detail of capillitium, and spores, × 250
- **d** Spore, × 1000

271 *Physarum mutabile* (Rost.) G. Lister
- **a** Two sporangia, × 5
- **b** Plasmodiocarp (after Lister), × 5
- **c** Two sporangia, one showing columella, × 20
- **d** Detail of capillitium, and spores, × 250
- **e** Spore, × 1000

272 *Physarum nicaraguense* Macbr.
- **a** Two sporangia, with stalks of others, × 5
- **b** Sporangium, × 20
- **c** Detail of capillitium, and spores, × 250
- **d** Spore, × 1000

273 *Physarum notabile* Macbr.
- **a** Cluster of sporangia, × 5
- **b** Isolated sporangium, × 20
- **c** Cluster of sporangia, × 20
- **d** Detail of capillitium, and spores, × 250
- **e** Spore, × 1000

274 *Physarum nucleatum* Rex
- **a** Cluster of sporangia, × 5
- **b** Two sporangia, × 20
- **c** Detail of capillitium, and spores, × 250
- **d** Spore, × 1000

275 *Physarum nudum* Macbr.
- **a** Cluster of sporangia, × 5
- **b** Sporangium, × 40
- **c** Detail of capillitium, and spores, × 250
- **d** Spore, × 1000

276 *Physarum nutans* Pers.
- **a** Three sporangia, × 5
- **b** Two sporangia, × 20
- **c** Detail of capillitium, and spores, × 250
- **d** Spore, × 1000

277 *Physarum oblatum* Macbr.
- **a** Two sporangia, with stalk of another, × 5
- **b** Two sporangia, × 20
- **c** Flabellate sporangium with short stem from same fruiting as b, × 20
- **d** Detail of capillitium, and spores, × 250
- **e** Spore, × 1000

278 *Physarum ovisporum* G. Lister
- **a** Sporangium and small plasmodiocarps, × 5
- **b** Sporangium, × 20
- **c** Detail of capillitium, and spores, × 250
- **d** Spore, × 1000

Plate XXX

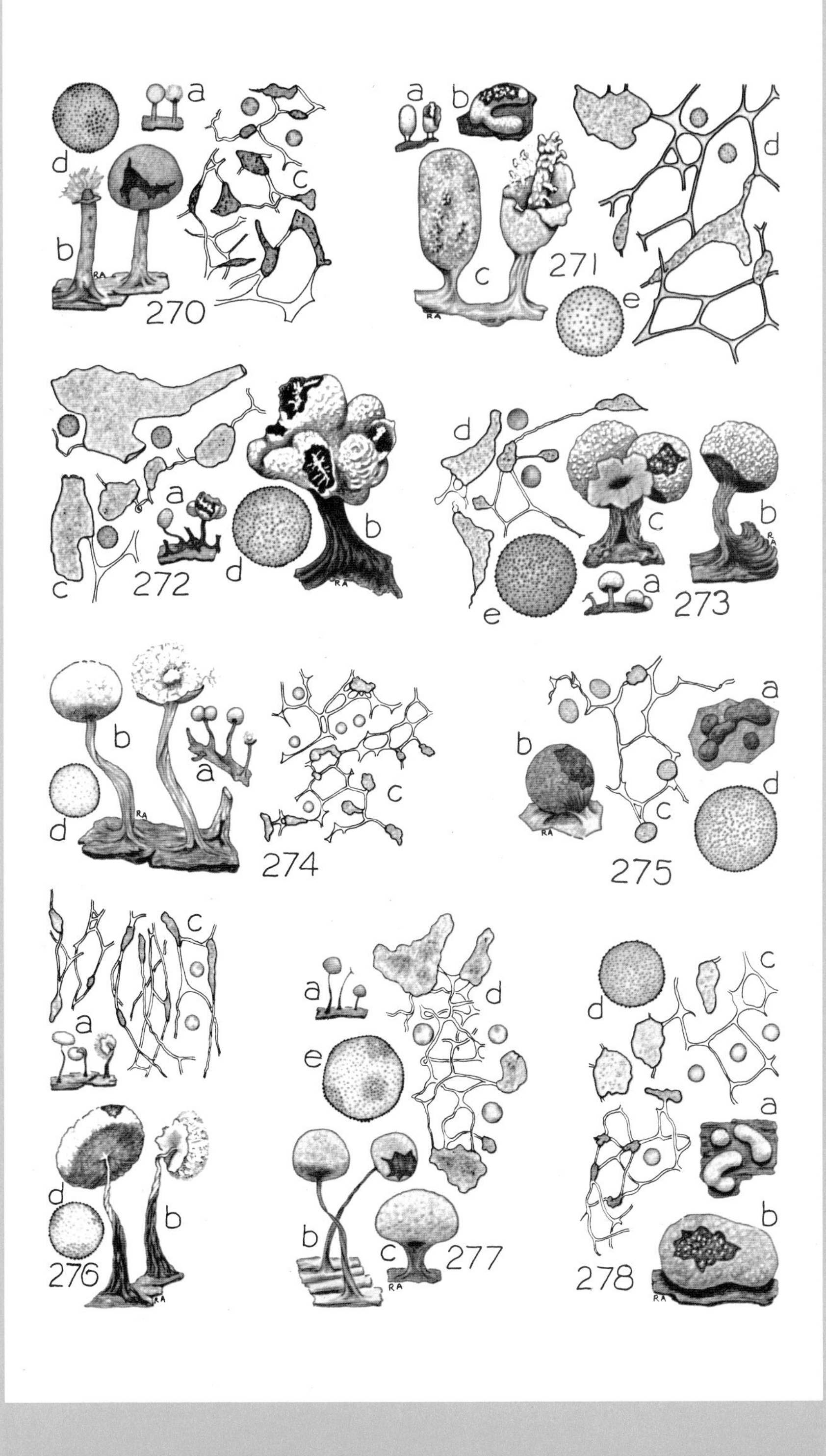

a
d
b
c
270
a
b
d
c
271
e
a
b
c
d
272
d
c
b
e
a
273
b
a
d
c
274
a
b
d
c
275
c
a
d
b
276
a
d
e
b
c
277
c
d
a
b
278

Plate XXXI

279 *Physarum penetrale* Rex
a Two sporangia, × 5
b Same, × 20
c Detail of capillitium, and spores, × 250
d Spore, × 1000

280 *Physarum pezizoideum* (Jungh.) Pav. & Lag. var. *pezizoideum*
a Two sporangia, × 5
b Same, × 20
c Detail of capillitium, showing attachment to base of peridium, and spores, × 250
d Spore, × 1000

281 *Physarum polycephalum* Schw.
a Two sporangia, × 5
b Same, × 20
c Detail of capillitium, and spores, × 250
d Spore, × 1000

282 *Physarum psittacinum* Ditmar
a Four sporangia, × 5
b/c Three sporangia, × 20
d Detail of capillitium, and spores, × 250
e Spore, × 1000

283 *Physarum pulcherrimum* Berk. & Rav.
a Two sporangia, × 5
b Two sporangia and remains of third, showing columella, × 20
c Detail of capillitium, and spores, × 250
d Spore, × 1000

284 *Physarum pulcherripes* Peck
a Two sporangia and remnant of third, showing columella, × 5
b Two sporangia, × 20
c Detail of capillitium, and spores, × 250
d Spore × 1000

285 *Physarum pusillum* (Berk. & Curt.) G. Lister
a Three sporangia, × 5
b Two sporangia, × 20
c Detail of capillitium, and spores from collection with physaroid capillitium, × 250
d Same, from collection with somewhat badhamioid capillitium, × 250
e Spore, × 1000

286 *Physarum retisporum* Martin, Thind & Rehill
a Two fructifications, × 5
b Sporangiate fruiting, × 20
c Plasmodiocarpous fruiting, × 20
d Detail of capillitium, and spores, × 250
e Spore, × 1000

Plate XXXI

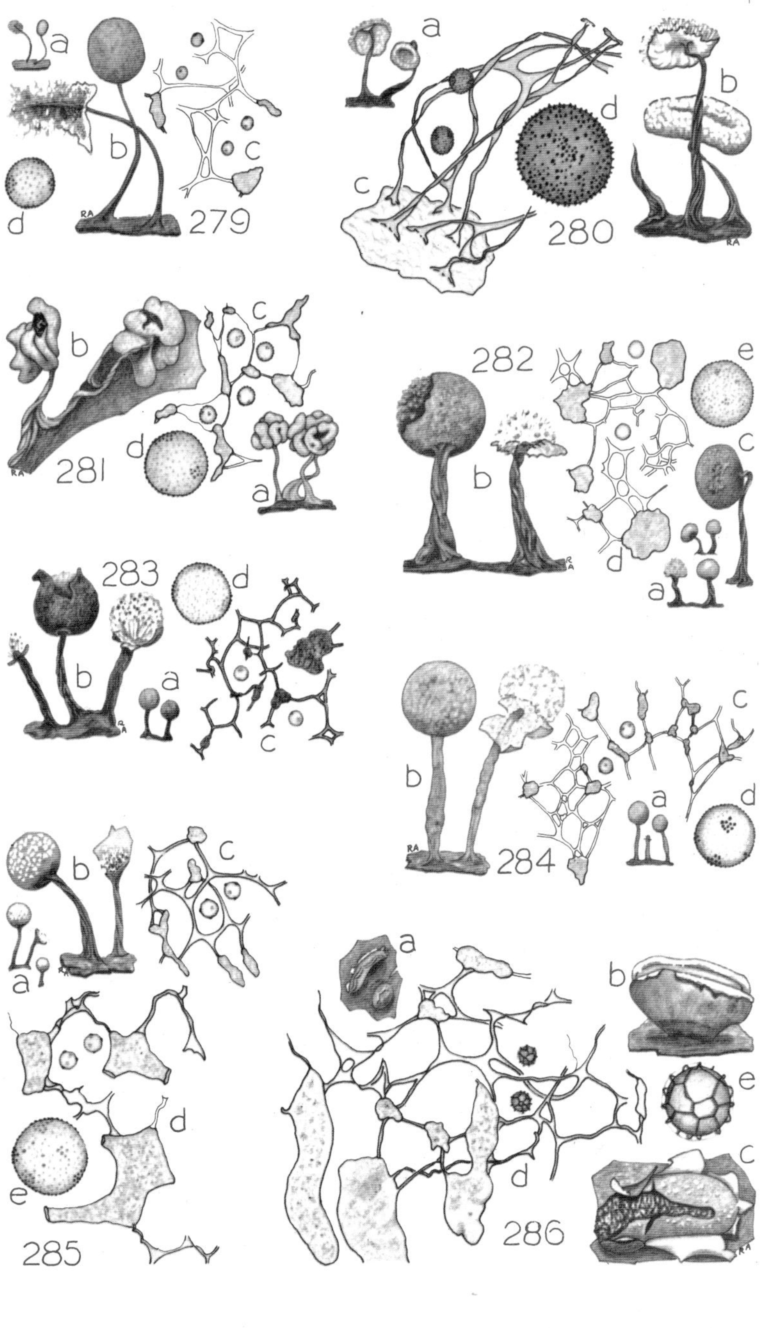

Plate XXXII

287 *Physarum roseum* Berk. & Br.
- **a** Six sporangia, × 5
- **b** Two sporangia, × 20
- **c** Detail of capillitium, and spores, × 250
- **d** Spore, × 1000

288 *Physarum rubiginosum* Fries
- **a** Cluster of brown sporangia, × 5
- **b** Two red sporangia, × 20
- **c** Detail of capillitium, and spores, × 250
- **d** Spore, × 1000

289 *Physarum rubronodum* Martin
- **a** Part of a cluster of sporangia, × 5
- **b** Sporangium, × 20
- **c** Detail of capillitium, and spores, × 250
- **d** Spore, × 1000

290 *Physarum serpula* Morgan
- **a** Plasmodicarps, × 5
- **b** Sporangiate fruiting, × 20
- **c** Detail of capillitium, and spores, × 250
- **d** Spore, × 1000

291 *Physarum spinulosum* Thind & Sehgal
- **a** Cluster of sporangia, × 5
- **b** Same, × 10
- **c** Detail of capillitium, and spores, × 250
- **d** Spore, × 1000

292 *Physarum stellatum* (Massee) Martin
- **a** Cluster of sporangia (after Lister), × 5
- **b** Two sporangia, × 20
- **c** Detail of capillitium, and spores, × 250
- **d** Spore, × 1000

293 *Physarum straminipes* A. Lister
- **a** Cluster of sporangia, × 5
- **b** Two sporangia, the stalks fused, × 20
- **c** Detail of capillitium, and spores, × 250
- **d** Spore, × 1000

294 *Physarum sulphureum* Alb. & Schw.
- **a** Two stalked sporangia, × 5
- **b** Sessile sporangium, × 5
- **c** Sporangium, × 20
- **d** Detail of capillitium, and spores, × 250
- **e** Spore, × 1000

Plate XXXII

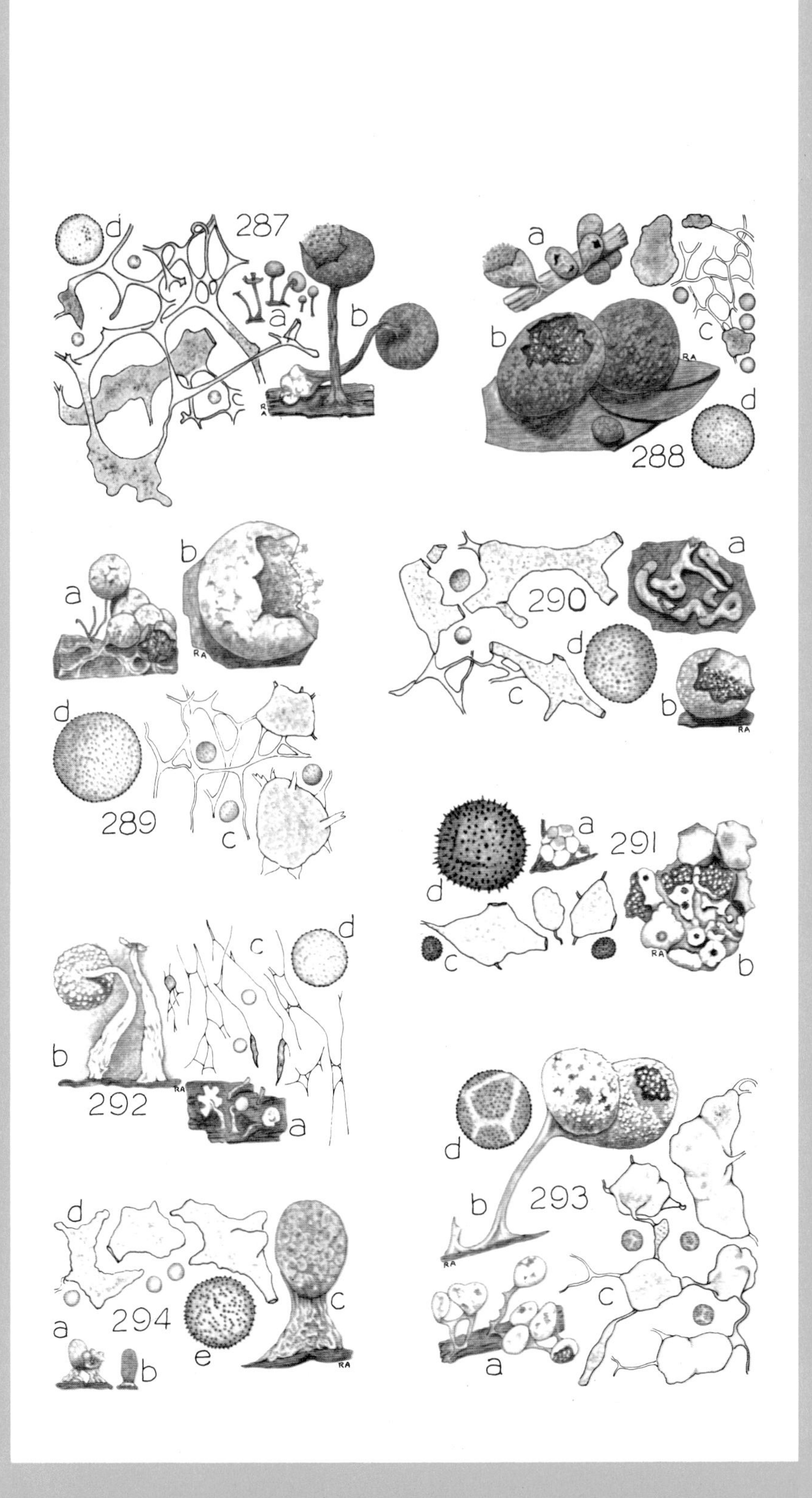

287
a
b
c
d
288
a
b
c
d
289
a
b
c
d
290
a
b
c
d
291
a
b
c
d
292
a
b
c
d
293
a
b
c
d
294
a
b
c
d
e

Plate XXXIII

295 *Physarum superbum* Hagelst.
- **a** Sporangium and plasmodiocarps, × 5
- **b** Plasmodiocarp, × 20
- **c** Detail of capillitium, and spores, × 250
- **d** Spore, × 1000

296 *Physarum tenerum* Rex
- **a** Sporangia, × 5
- **b** Two sporangia, × 20
- **c** Detail of capillitium, and spores, × 250
- **d** Spore, × 1000

297 *Physarum tessellatum* Martin & Farr
- **a** Cluster of sporangia, × 5
- **b** Portion of same, × 20
- **c** Detail of capillitium, and spores, × 250
- **d** Spore, × 1000

298 *Physarum tropicale* Macbr.
- **a** Sporangium and stalk of another, × 5
- **b** Sporangium, × 20
- **c** Detail of capillitium, and spores, × 250
- **d** Spore, × 1000

299 *Physarum vernum* Somm.
- **a** Two clustered fructifications, × 5
- **b** Detail of capillitium, and spores, × 250
- **c** Spore, × 1000

300 *Physarum virescens* Ditmar
- **a** Cluster of sporangia, × 5
- **b** Same, × 20
- **c** Detail of capillitium, and spores, × 250
- **d** Spore, × 1000

301 *Physarum viride* (Bull.) Pers.
- **a** Four sporangia, × 5
- **b** Yellow sporangium with broken peridium, × 20
- **c** Paler sporangium, × 20
- **d** White sporangium showing yellow nodes
- **e** Detail of capillitium, and spores, × 250
- **f** Spore, × 1000

302 *Wilczekia evelinae* Meylan
- **a** Two sporangia, × 5
- **b** Sporangium, × 20
- **c** Detail of capillitium, and spores, × 250
- **d** Spore, × 1000

303 *Physarina echinospora* Thind & Manocha
- **a** Two sporangia, × 5
- **b** Sporangium, × 20
- **c** Detail of capillitium, and spores, × 250
- **d** Spore, × 1000

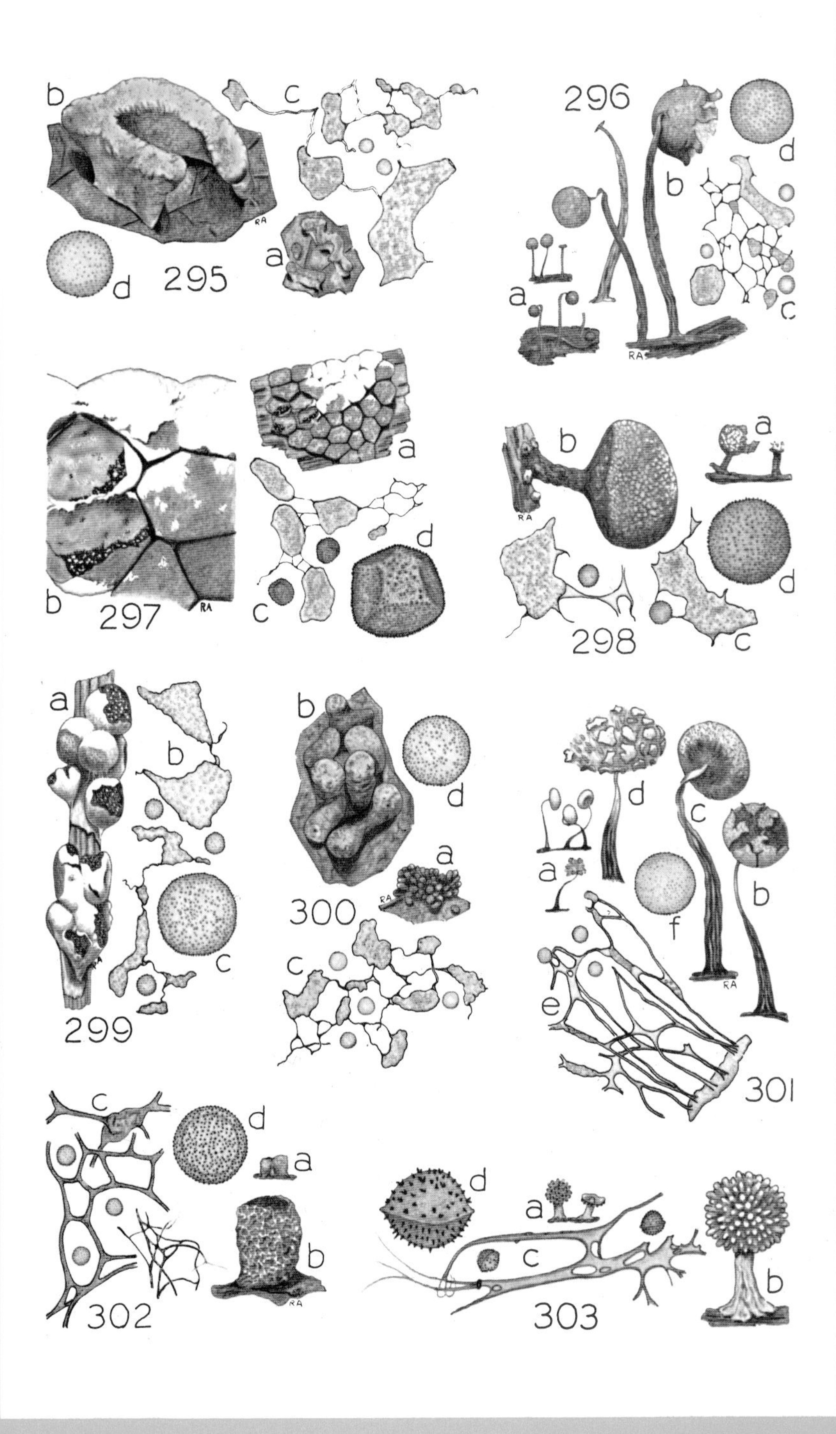
295
b
c
a
d
296
a
b
c
d
297
a
b
c
d
298
a
b
c
d
299
a
b
c
300
a
b
c
d
301
a
b
c
d
e
f
302
a
b
c
d
303
a
b
c
d

Plate XXXIV

304 *Diderma alpinum* Meylan
- **a** Part of a large cluster of fruitings, × 5
- **b** Detail of capillitium, calcareous body, and spores, × 250
- **c** Spore, × 1000

305 *Diderma asteroides* (A. & G. Lister) G. Lister
- **a** Group of sporangia, × 5
- **b** Open sporangium, × 20
- **c** Detail of capillitium, with spores, × 250
- **d** Spore, × 1000

306 *Diderma chondrioderma* (de Bary & Rost.) G. Lister
- **a** Cluster of sporangia, × 5
- **b** Same, × 20
- **c** Detail of capillitium, with tips attached to peridium, and spores, × 250
- **d** Spore, × 1000

307 *Diderma cinereum* Morgan
- **a** Cluster of sporangia, × 5
- **b** Same, × 20
- **c** Detail of capillitium, and spores, × 250
- **d** Spore, × 1000

308 *Diderma cor-rubrum* Macbr.
- **a** Two sporangia, with hypothallus, × 5
- **b** Three sporangia, one showing columella, × 20
- **c** Detail of capillitium, with spores and crystalline bodies, × 250
- **d** Spore, × 1000

309 *Diderma crustaceum* Peck
- **a** Part of an extensive fruiting, × 5
- **b** Portion of same, × 20
- **c** Detail of capillitium, and spores, × 250
- **d** Spore, × 1000

310 *Diderma darjeelingense* Thind and Sehgal
- **a** Three sporangia, × 5
- **b/c** Three sporangia, × 20
- **d** Detail of capillitium, and spores, × 250
- **e** Spore, × 1000

311 *Diderma deplanatum* Fries
- **a** Plasmodiocarp, × 5
- **b** Annulate plasmodiocarp, showing inner wall, × 20
- **c** Detail of capillitium, and spores, × 250
- **d** Spore, × 1000

312 *Diderma effusum* (Schw.) Morgan
- **a** Part of a large sporangiate fruiting, × 5
- **b** Plasmodiocarp, × 5
- **c** Mass of small, almost merged sporangia, × 5
- **d** Detail of capillitium, and spores, × 250
- **e** Spore, × 1000

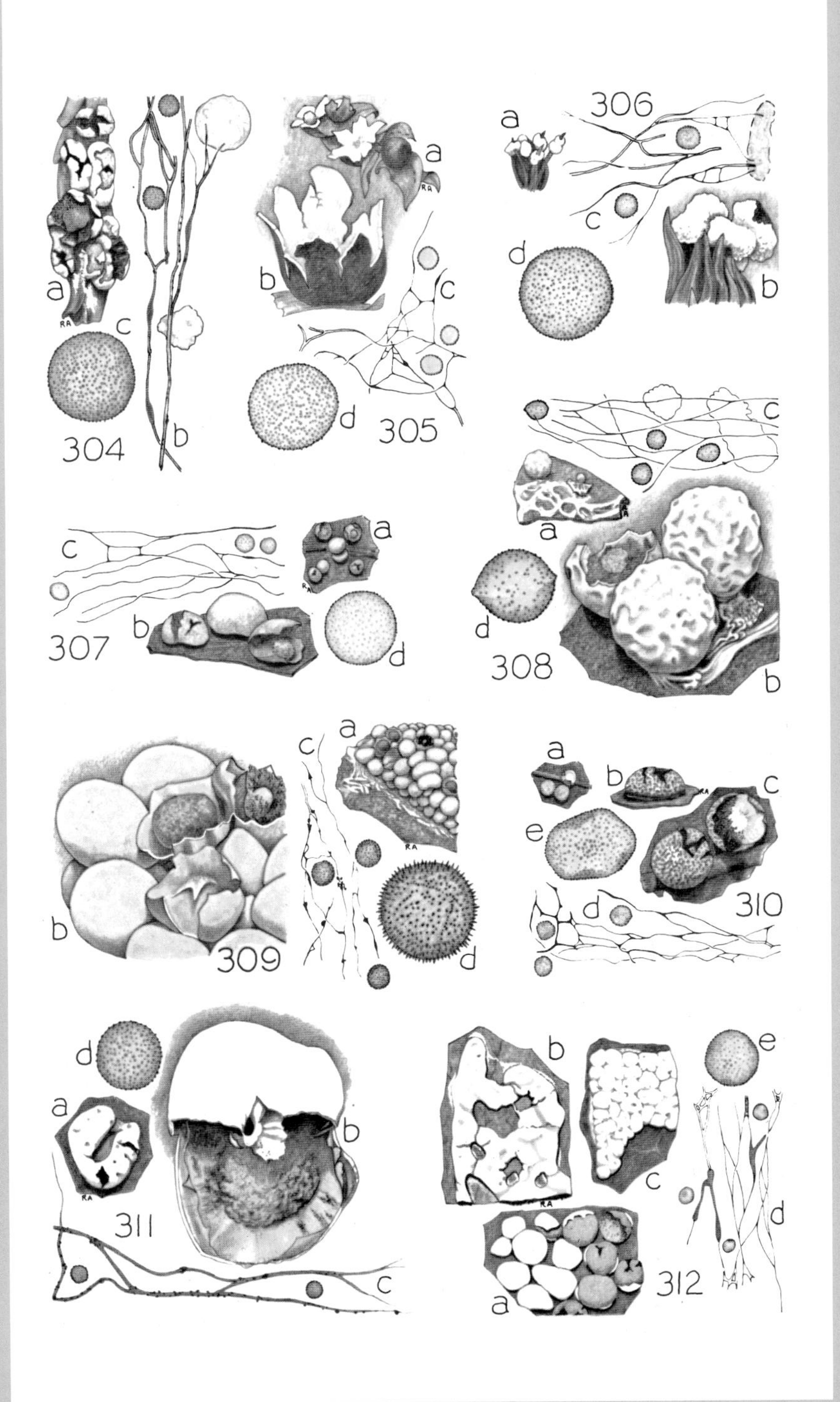
304
a
b
c
305
a
b
c
d
306
a
b
c
d
307
a
b
c
d
308
a
b
c
d
309
a
b
c
d
310
a
b
c
d
e
311
a
b
c
d
312
a
b
c
d
e

Plate XXXV

313 *Diderma floriforme* (Bull.) Pers.
- **a** Cluster of sporangia, × 5
- **b** Two sporangia, × 20
- **c** Detail of capillitium, and spores, × 250
- **d** Spore, × 1000

314 *Diderma globosum* Pers.
- **a** Group of sporangia, × 5
- **b** Three sporangia, × 20
- **c** Detail of capillitium, and spores, × 250
- **d** Spore, × 1000

315 *Diderma hemisphaericum* (Bull.) Hornem.
- **a** Group of sporangia, × 5
- **b** Sporangium, × 20
- **c** Detail of capillitium, and spores, × 250
- **d** Spore, × 1000

316 *Diderma indicum* Thind & Sehgal
- **a** Cluster of sporangia, × 5
- **b** Detail of capillitium, and spores, × 250
- **c** Spore, × 1000

317 *Diderma lucidum* Berk. & Br.
- **a** Sporangium, × 5
- **b** Two sporangia, × 20
- **c** Detail of capillitium showing attachment to fragment of peridium, and spores, × 250
- **d** Spore, × 1000

318 *Diderma lyallii* (Massee) Macbr.
- **a** Group of sporangia, × 5
- **b** Sporangium, × 20
- **c** Detail of capillitium, and spores, × 250
- **d** Spore, × 1000

319 *Diderma montanum* (Meylan) Meylan
- **a** Three sporangia, × 5
- **b** Sporangium, × 20
- **c** Detail of capillitium, and spores, × 250
- **d** Spore, × 1000

320 *Diderma mussooriense* Thind & Manocha
- **a** Cluster of sporangia, × 5
- **b** Two sporangia, × 20
- **c** Detail of capillitium, and spores, × 250
- **d** Spore, × 1000

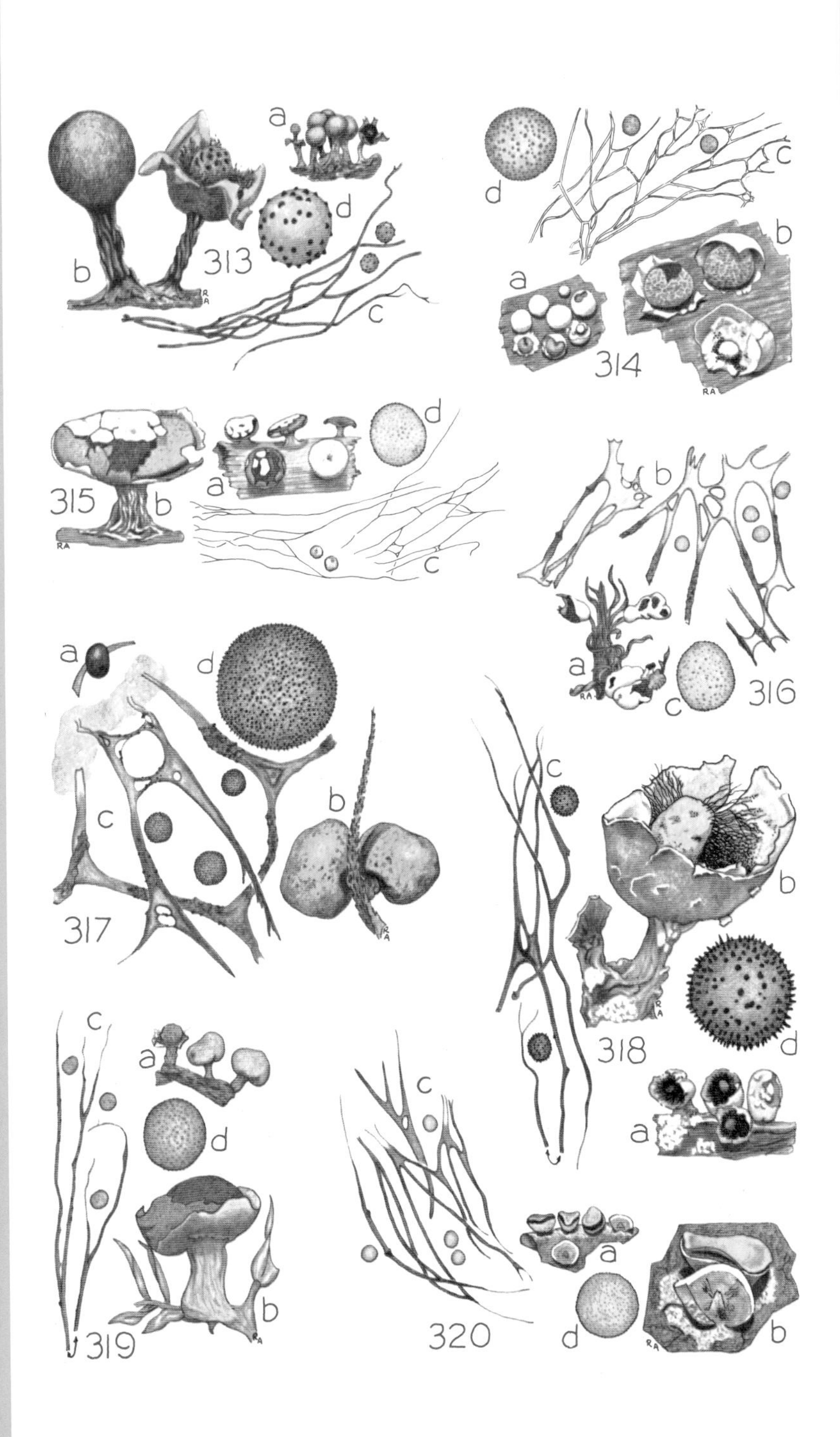
313
a
b
c
d
314
a
b
c
d
315
a
b
c
d
316
a
b
c
317
a
b
c
d
318
a
b
c
d
319
a
b
c
d
320
a
b
c
d

Plate XXXVI

321 *Diderma niveum* (Rost.) Macbr.
- **a** Cluster of sporangia, × 5
- **b** Detail of capillitium, and spores, × 250
- **c** Spore, × 1000

322 *Diderma ochraceum* Hoffm.
- **a** Three sporangia, × 5
- **b** Sporangium, × 20
- **c** Details of capillitium, and spores, × 250
- **d** Spore, × 1000

323 *Diderma radiatum* (L.) Morgan
- **a** Cluster of sporangia, × 5
- **b** Weathered sporangium, × 10
- **c** Detail of capillitium, and spores, × 250
- **d** Spore, × 1000

324 *Diderma roanense* (Rex) Macbr.
- **a** Three sporangia, × 5
- **b** Sporangium, × 10
- **c** Detail of capillitium, and spores, × 250
- **d** Spore, × 1000

325 *Diderma rugosum* (Rex) Macbr.
- **a** Two sporangia, × 5
- **b** Two sporangia, × 20
- **c** Detail of capillitium, and spores, × 250
- **d** Spore, × 1000

326 *Diderma simplex* (Schroet.) G. Lister
- **a** Group of sporangia, × 5
- **b** Sporangium, × 20
- **c** Detail of capillitium, and spores, × 250
- **d** Spore, × 1000

327 *Diderma spumarioides* (Fries) Fries
- **a** Part of a large cluster of sporangia, × 5
- **b** Sporangium cut through center to show columella, with base of old sporangium, × 20
- **c** Detail of capillitium, and spores, × 250
- **d** Spore, × 1000

328 *Diderma subdictyosporum* (Rost.) G. Lister
- **a** Details of capillitium, and spores, × 250
- **b** Spore, × 1000

329 *Diderma testaceum* (Schrad.) Pers.
- **a** Group of sporangia, × 5
- **b** Sporangium, × 20
- **c** Detail of capillitium, and spores, × 250
- **d** Spore, × 1000

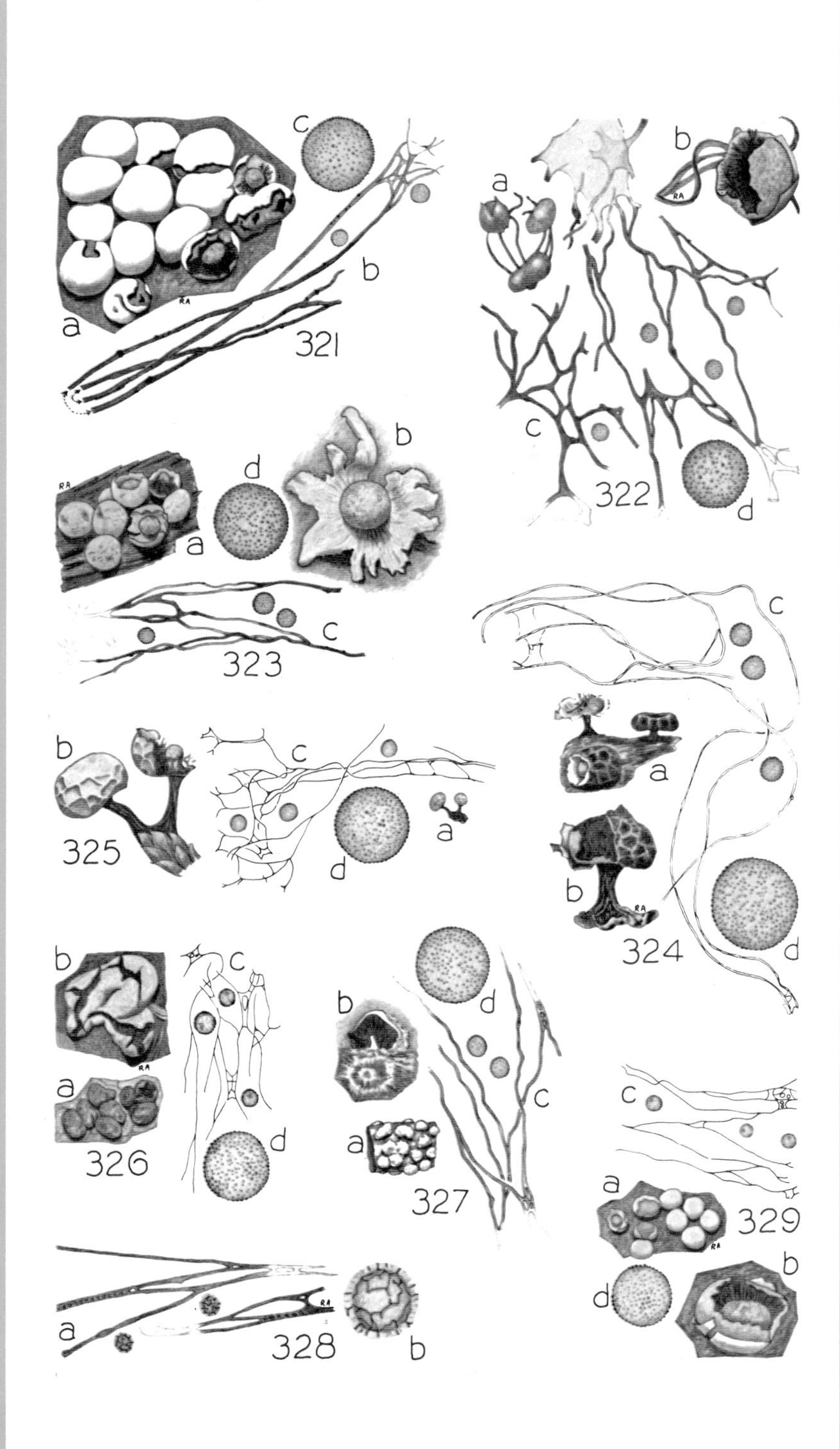
321
a
b
c
322
a
b
c
d
323
a
c
d
b
324
a
b
c
d
325
a
b
c
d
326
a
b
c
d
327
a
b
c
d
328
a
b
329
a
b
c
d

Plate XXXVII

330 *Diderma travelyani* (Grev.) Fries
- **a** Cluster of sporangia, × 5
- **b** Two sporangia, one before opening, the other after spore discharge, × 10
- **c** Details of capillitium, portion of peridium with imbedded lime crystals, and spores, × 250
- **d** Spore, × 1000

331 *Mucilago crustacea* Wiggers
- **a** Small aethalium, × ½
- **b** Detail, showing portions of capillitium and pseudocapillitium, with crystals and spores, × 250
- **c** Spore, × 1000

332 *Didymium anellus* Morgan
- **a** Cluster of fructifications, × 5
- **b** Plasmodiocarp, × 20
- **c** Detail of capillitium, with crystals and spores, × 250
- **d** Spore, × 1000

333 *Didymium clavus* (Alb. & Schw.) Rab.
- **a** Two sporangia, ×5
- **b** Three sporangia, × 20
- **c** Detail of capillitium, with crystals and spores, × 250
- **d** Spore, × 1000

334 *Didymium crustaceum* Fries
- **a** Massed sporangia ,× 5
- **b** Sporangia, one with stalk-like hypothallus, × 10
- **c** Detail of capillitium, with crystals and spores, × 250
- **d** Spore, × 1000

335 *Didymium decipiens* Meylan
- **a** Portions of capillitium, and spores, × 250
- **b** Spore, × 1000

336 *Didymium difforme* (Pers.) S. F. Gray
- **a** Cluster of fruitings, × 5
- **b** Plasmodiocarp, × 20
- **c** Detail of capillitium, with crystals and spores, × 250
- **d** Spore, × 1000

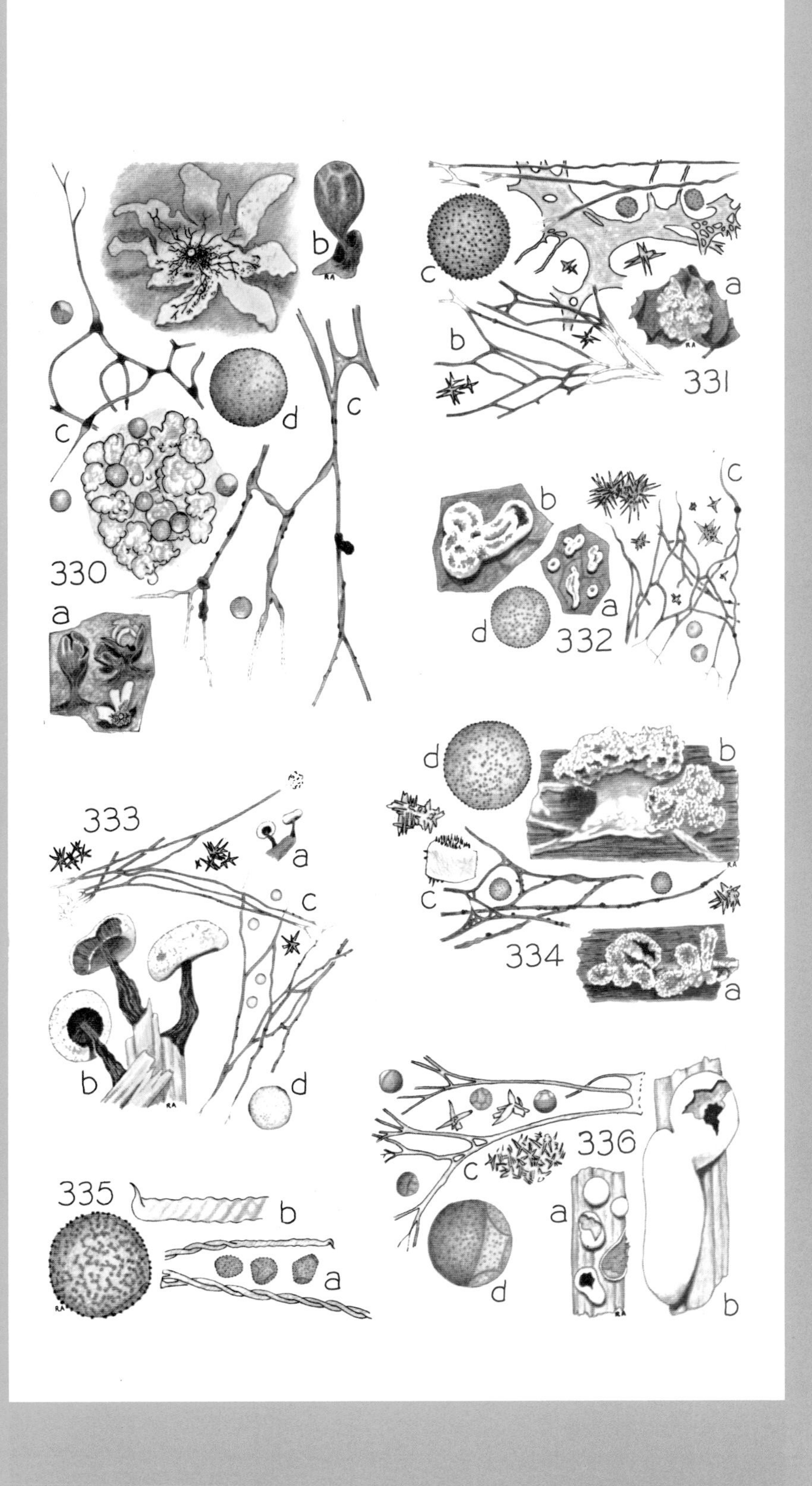
330
a
b
c
c
d
331
a
b
c
332
a
b
c
d
333
a
b
c
d
334
a
b
c
d
335
a
b
336
a
b
c
d

Plate XXXVIII

337 *Didymium dubium* Rost.
 a Plasmodiocarp, × 5
 b Detail of capillitium, with spores and crystals, × 250
 c Spore, × 1000

338 *Didymium megalosporum* Berk. & Curt.
 a Three sporangia, × 5
 b Two sporangia, one showing internal structure, × 20
 c Detail of capillitium showing attachment to peridium, with spores and crystals, × 250
 d Spore, × 1000

339 *Didymium flexuosum* Yamashiro
 a Portion of plasmodiocarp, × 5
 b Enlarged detail of same, × 20
 c Detail of capillitium, with vesicular bodies, spores and crystals, × 250
 d Spore, × 1000

340 *Didymium floccosum* Martin, Thind & Rehill
 a Two sporangia, × 5
 b Two sporangia with columella of a third, × 20
 c Detail of capillitium showing attachment to peridium, with crystals and spores, × 250
 d Spore, × 1000

341 *Didymium fulvum* Sturgis
 a Plasmodiocarp, × 5
 b Detail of capillitium, with spores and crystals, × 250
 c Spore, × 1000

342 *Didymium intermedium* Schroet.
 a Group of sporangia, × 5
 b Two sporangia, that on right showing deep umbilicus, × 20
 c Detail of capillitium, with spores, crystals, and part of stalk with included crystals, × 250
 d Spore, × 1000

343 *Didymium iridis* (Ditmar) Fries
 a Group of sporangia, × 5
 b Sporangium, × 20
 c Detail of capillitium, with spores and crystals, × 250
 d Spore, × 1000

344 *Didymium leoninum* Berk. & Br.
 a Sporangium, × 5
 b Two sporangia, × 20
 c Detail of capillitium, showing attachment to peridium, with spores and crystals, × 250
 d Spore, × 1000

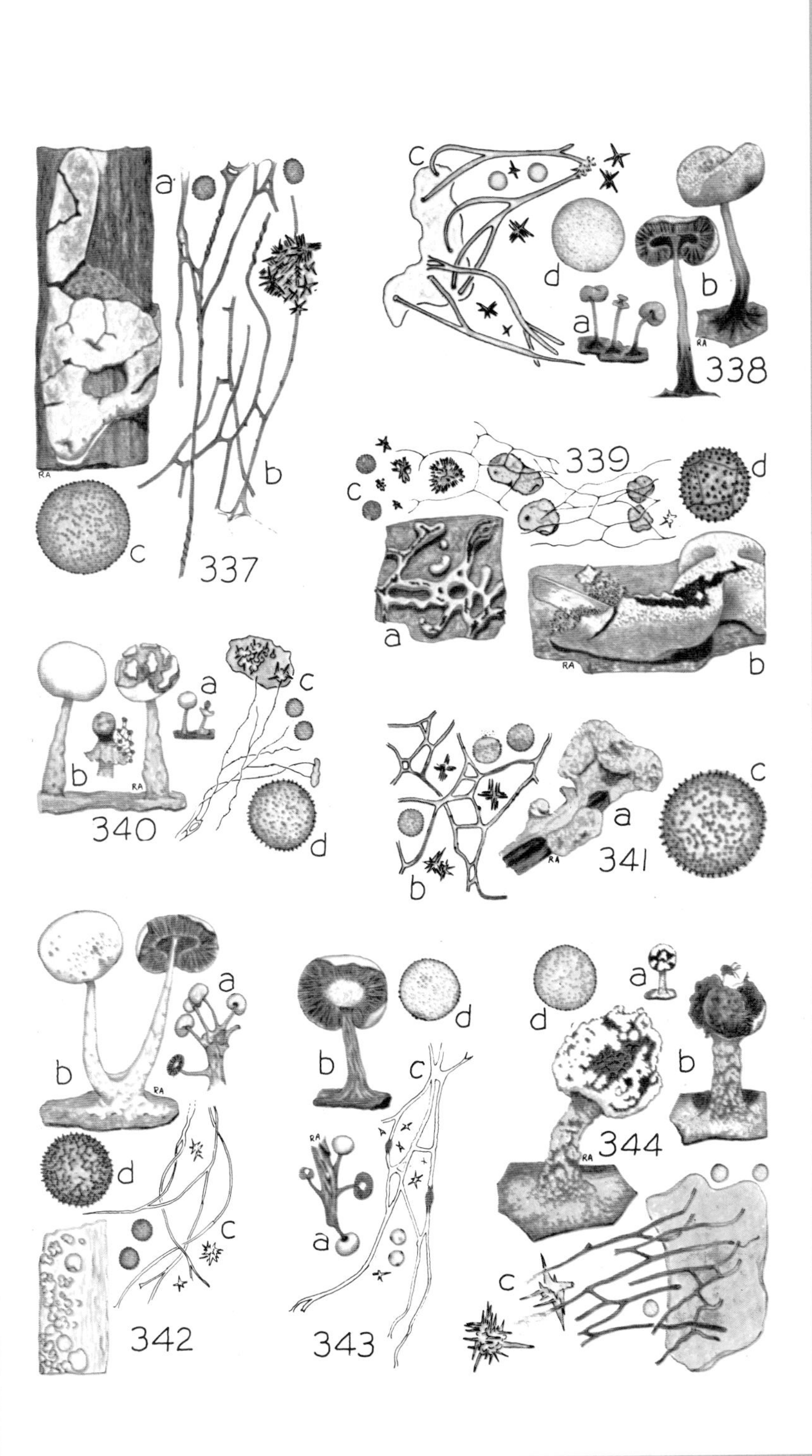
337
a
b
c
338
a
b
c
d
339
a
b
c
d
340
a
b
c
d
341
a
b
c
342
a
b
c
d
343
a
b
c
d
344
a
b
c
d

Plate XXXIX

345 *Didymium listeri* Massee
- **a** Small plasmodiocarp, × 5
- **b** Sporangium, × 20
- **c** Details of capillitium, with spores and crystals, × 250
- **d** Spore, × 1000

346 *Didymium melanospermum* (Pers.) Macbr.
- **a** Two sporangia, × 5
- **b** Same, × 20
- **c** Portion of peridium, showing attached capillitium, ×50
- **d** Detail of same, with spores and crystals, × 250
- **e** Spore, × 1000

347 *Didymium minus* (A. Lister) Morgan
- **a** Sporangium, × 5
- **b** Two sporangia, × 20
- **c** Details of capillitium, showing attachment to peridium, with crystals and spores, × 250
- **d** Spore, × 1000

348 *Didymium nigripes* (Link) Fries
- **a** Sporangium, × 5
- **b** Two sporangia, × 20
- **c** Details of capillitium, showing attachment to peridium, with crystals and spores, × 250
- **d** Spore, × 1000

349 *Didymium ochroideum* G. Lister
- **a** Three sporangia, × 5
- **b** Five sporangia, × 20
- **c** Detail of capillitium, with spores and crystals, × 250
- **d** Spore, × 1000

350 *Didymium ovoideum* Nann.-Brem.
- **a** Sporangium, × 5
- **b** Two sporangia, × 20
- **c** Empty sporangium showing rugose columella, × 20
- **d** Detail of capillitium, showing enlargements and attachment to peridium, with crystals and spores, × 250
- **e** Spore, × 1000

351 *Didymium perforatum* Yamashiro
- **a** Portion of plasmodiocarp, × 5
- **b** Detail of same, × 20
- **c** Detail of capillitium, with crystals and spores, × 250
- **d** Spore, × 1000

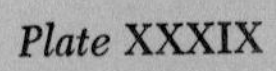

Plate XXXIX

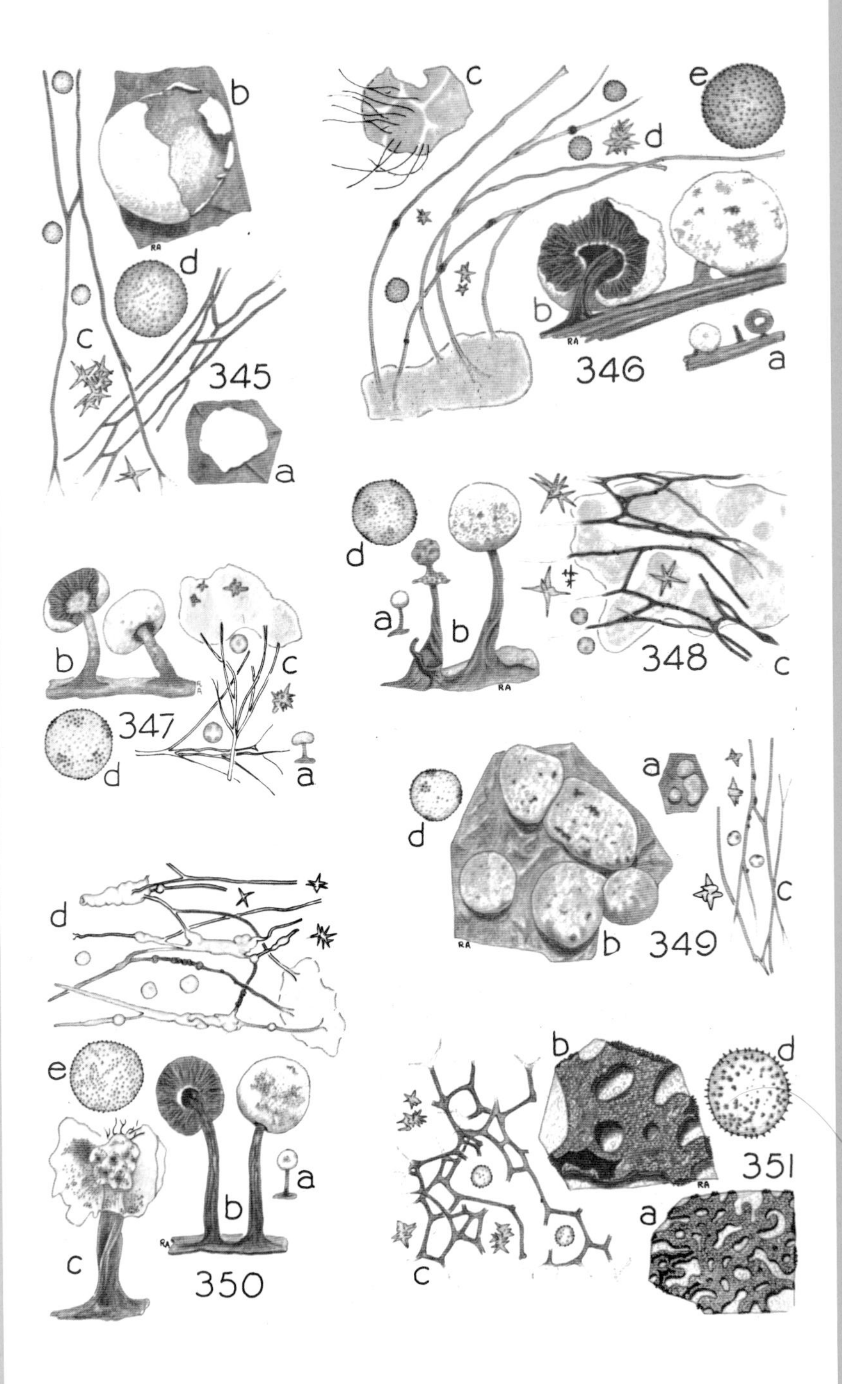
345
a
b
c
d
346
a
b
c
d
e
347
a
b
c
d
348
a
b
c
d
349
a
b
c
d
350
a
b
c
d
e
351
a
b
c
d

Plate XL

352 *Didymium quitense* (Pat.) Torrend
- **a** Three fructifications, × 5
- **b** Detail of capillitium, with crystals and spores, × 250
- **c** Spore, × 1000

353 *Didymium serpula* Fries
- **a** Plasmodiocarps, × 5
- **b** Detail of plasmodiocarp, × 20
- **c** Details of capillitium, with vesicles, crystals and spores, × 250
- **d** Spore, × 1000

354 *Didymium squamulosum* (Alb. & Schw.) Fries
- **a** Sporangia, × 5
- **b/c** Sporangia, × 20
- **d** Detail of capillitium showing attachment to peridium, with crystals and spores, × 250
- **e** Spore, × 1000

355 *Didymium sturgisii* Hagelst.
- **a** Plasmodiocarp, × 5
- **b** Detail of same, × 20
- **c** Diagram of same, showing trabeculae, × 20
- **d** Detail of capillitium showing attachment to peridium, with trabeculae, crystals and spores, × 250
- **e** Spore, × 1000

356 *Didymium trachysporum* G. Lister
- **a** Cluster of sporangia, × 5
- **b** Three sporangia, × 20
- **c** Details of capillitium, with fragment of peridium, crystals and spores, × 250
- **d** Spore, × 1000

357 *Didymium vaccinum* (Dur. & Mont.) Buchet
- **a** Cluster of fructifications, × 5
- **b** Sporangium, × 20
- **c** Detail of capillitium, showing attachment to columella, with crystals and spores, × 250
- **d** Spore, × 1000 (California)
- **e** Spore, × 1000 (England)

358 *Didymium verrucosporum* Welden
- **a** Sporangia, × 5
- **b** Two sporangia, × 20
- **c** Detail of capillitium, with attached peridial fragment, and crystals and spores, × 250
- **d** Spore, × 1000

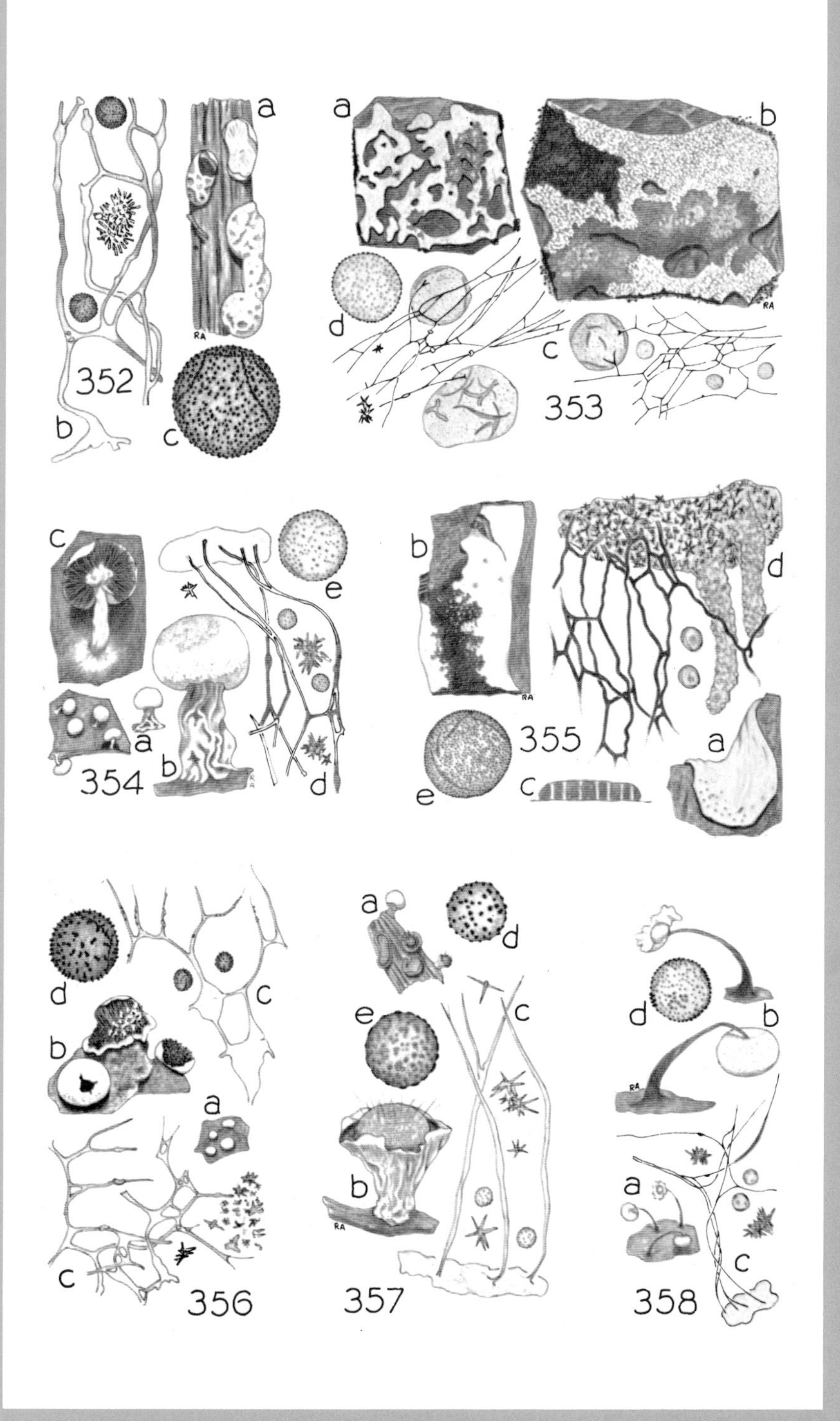
352
a
b
c
353
a
b
c
d
354
a
b
c
d
e
355
a
b
c
d
e
356
a
b
c
d
357
a
b
c
d
e
358
a
b
c
d

Plate XLI

359 *Lepidoderma carestianum* (Rab.) Rost.
- **a** Sporangiate and plasmodiocarpous fructifications, × 5
- **b** Capillitium and spores, × 250
- **c** Spore, × 1000

360 *Lepidoderma chailletii* Rost.
- **a** Cluster of sporangia, × 5
- **b** Sporangium, × 20
- **c** Capillitium, cluster of crystals from scale, and spores, × 250
- **d** Spore, × 1000

361 *Lepidoderma granuliferum* (Phill.) R. E. Fries
- **a** Fructifications, × 5
- **b** Capillitium, with vesicles containing crystals, and spores, × 250
- **c** Spore, × 1000

362 *Lepidoderma tigrinum* (Schrad.) Rost.
- **a** Sporangium, × 5
- **b** Sporangium, × 10
- **c** Capillitium, scales, and spores, × 250
- **d** Spore, × 1000

363 *Leptoderma iridescens* G. Lister
- **a** Three sporangia, two partly fused, × 5
- **b** Four sporangia, × 20
- **c** Capillitium, showing attachment to peridium, and spores, × 250
- **d** Spore, × 1000

364 *Didymium laxifila* G. Lister & Ross
- **a** Cluster of sporangia, × 5
- **b** Sporangium, × 20
- **c** Capillitium, crystals, and spores, × 250
- **d** Spore, × 1000

365 *Licea pumila* Martin & Allen
- **a** Cluster of sporangia, × 5
- **b** Two sporangia, × 50
- **c** Spore, × 1000

366 *Badhamia ainoae* Yamashiro
- **a** Three sporangia, × 5
- **b** Sporangium, partly open, to show limy columns, × 20
- **c** Single column, with fragments of base and peridium attached, and spores, × 50
- **d** Spore, × 1000

367 *Hemitrichia chrysospora* (A. Lister) A. Lister
- **a** Sporangia and plasmodiocarps, × 10
- **b** Capillitium and spores, × 500
- **c** Spore, × 1000

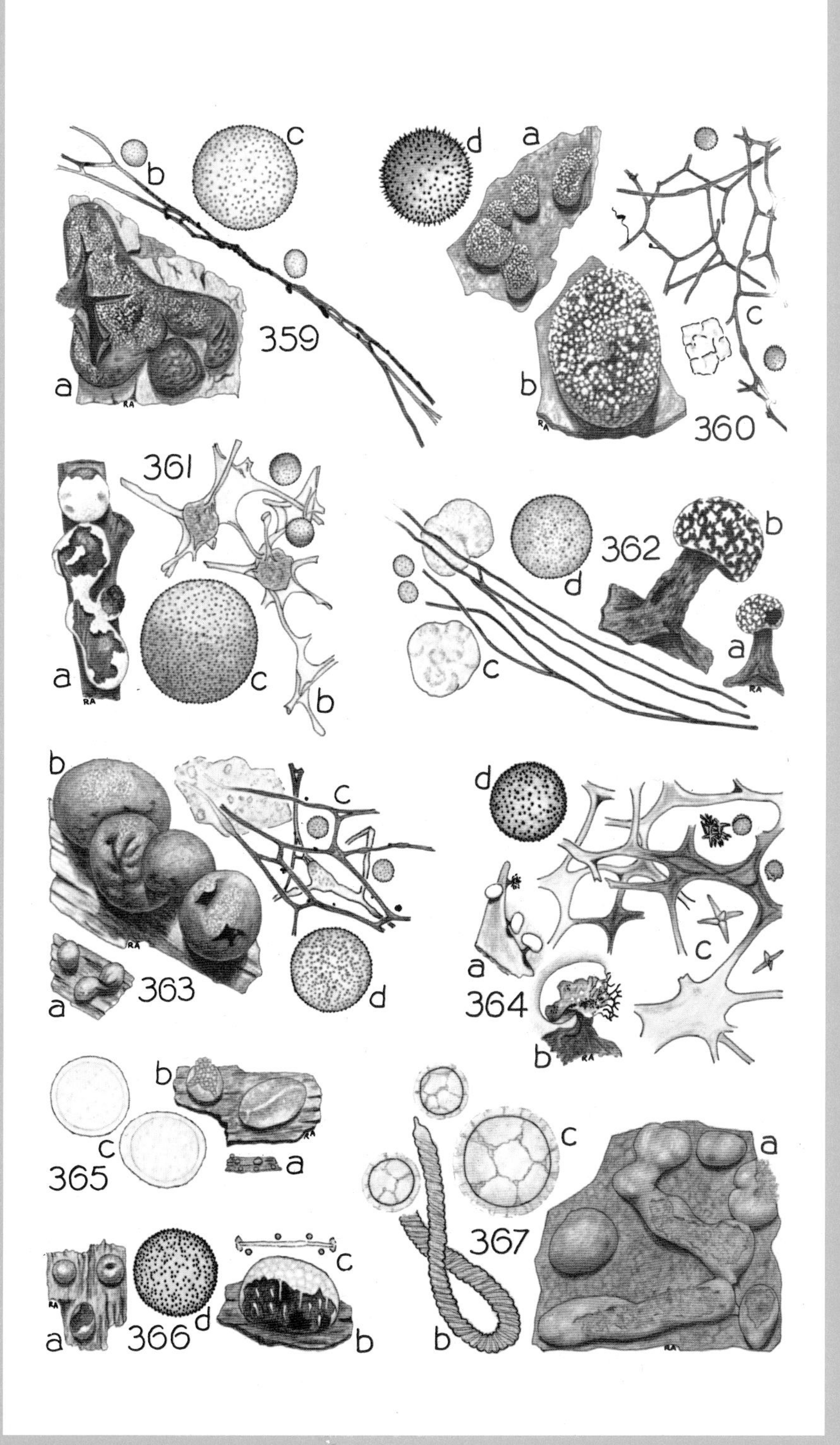
359
a
b
c
d
360
361
362
363
364
365
366
367